Pro-Poor Mountain Tourism

This timely and interdisciplinary book is the first to examine mountain tourism and local communities with a pro-poor lens. By drawing on human geography, political and social science, ethics and moral philosophy and empirical research, the volume explores how mountain tourism can be used to fight poverty and inequality in mountain regions.

Mountain tourism represents a growing mass tourism phenomenon. The local population, recognizing the possibilities for increased income, started to develop in situ services. However, sensitive to outside influences, the environment of high-altitude mountain areas resident communities have been abruptly exposed to impacts from mountain tourism-related activities, although until recently, they have been cut off from civilization. The natural environment and people living in mountain regions have been affected by an increasing number of visitors in the last few decades. Hence, this book provides an expert-led and comprehensive summary of mountain tourism development and illustrates how tourism can increase benefits for the poor within local communities. Furthermore, it presents updated management and governance policies.

This volume will be of pivotal interest to scholars and practitioners from the fields of geography and tourism studies, ethics, and development economics, as well as policymakers, aid agencies, and general readers interested in sustainable development in mountain regions.

Michal Apollo is an Assistant Professor at the Institute of Earth Sciences, the University of Silesia in Katowice, Poland, and a Fellow of Yale University's Global Justice Program, New Haven, USA. Michal is an enthusiastic researcher, traveller, mountaineer, ultra-runner, sailor, diver, photographer, and science populariser. Michal's unique background allows him to integrate knowledge from the perspectives of various points of view into his research and consultancy work. His areas of expertise are tourism management, consumer behaviours well as environmental and socio-economic issues. In his main research field, he focuses on human presence in high mountain regions and their wellbeing. Currently, he is working on a concept of sustainable use of environmental and human resources, as this is a key to the development, prosperity and wellbeing of all stakeholders. Michal is a board member of the Academics Stand Against Poverty.

Yana Wengel is an Associate Professor at the Hainan University – Arizona State University International Tourism College based in Hainan University, China. Yana takes a critical approach to tourism studies, and her interests include volunteer tourism, non-profit tourism, tourism in developing economies, creative methodologies and mountain tourism. She has an interest in creative qualitative tools for data collection and stakeholder engagement. Yana is a co-founder of the LEGO® SERIOUS PLAY® research community.

Thomas Pogge is a Leitner Professor of Philosophy and International Affairs and founding Director of the Global Justice Program at Yale University, USA. Thomas Pogge is a member of the Norwegian Academy of Science, as well as co-founder of Academics, Stand Against Poverty (ASAP), an international network aiming to enhance the impact of scholars, teachers and students on global poverty, and of Incentives for Global Health, a team effort toward developing a complement to the pharmaceutical patent regime that would improve access to advanced medicines for the poor worldwide. Pogge has published widely on Immanuel Kant and in moral and political philosophy, including various books on Rawls and global justice.

Contemporary Geographies of Leisure, Tourism and Mobility

Series Editor: C. Michael Hall, *Professor at the Department of Management, College of Business and Economics, University of Canterbury, Christchurch, New Zealand*

The aim of this series is to explore and communicate the intersections and relationships between leisure, tourism and human mobility within the social sciences.

It will incorporate both traditional and new perspectives on leisure and tourism from contemporary geography, e.g. notions of identity, representation and culture, while also providing for perspectives from cognate areas such as anthropology, cultural studies, gastronomy and food studies, marketing, policy studies and political economy, regional and urban planning, and sociology, within the development of an integrated field of leisure and tourism studies.

Also, increasingly, tourism and leisure are regarded as steps in a continuum of human mobility. Inclusion of mobility in the series offers the prospect to examine the relationship between tourism and migration, the sojourner, educational travel, and second home and retirement travel phenomena.

The series comprises two strands:

Contemporary Geographies of Leisure, Tourism and Mobility aims to address the needs of students and academics, and the titles will be published simultaneously in hardback and paperback.

Tourism and Innovation 2nd Edition
C. Michael Hall and Allan M. Williams

Tourism and Animal Ethics 2nd Edition
David A. Fennell

Routledge Studies in Contemporary Geographies of Leisure, Tourism and Mobility is a forum for innovative new research intended for research students and academics, and the titles will initially be available in hardback only.

Pro-Poor Mountain Tourism
Edited by Michal Apollo, Yana Wengel and Thomas Pogge

Tourism Safety, Security and Resilience
Integrated Community-Based Approaches
Edited by Rami K. Isaac, Shem Wambugu Maingi and Vanessa GB Gowresunkaar

For more information about this series, please visit: www.routledge.com/Contemporary-Geographies-of-Leisure-Tourism-and-Mobility/book-series/SE0522

Pro-Poor Mountain Tourism

Edited by
Michal Apollo, Yana Wengel
and Thomas Pogge

Routledge
Taylor & Francis Group
LONDON AND NEW YORK

First published 2025
by Routledge
4 Park Square, Milton Park, Abingdon, Oxon OX14 4RN

and by Routledge
605 Third Avenue, New York, NY 10158

Routledge is an imprint of the Taylor & Francis Group, an informa business

British Library Cataloguing-in-Publication Data
A catalogue record for this book is available from the British Library

Library of Congress Cataloging-in-Publication Data
A catalog record has been requested for this book

ISBN: 978-1-032-49199-8 (hbk)
ISBN: 978-1-032-49200-1 (pbk)
ISBN: 978-1-003-39258-3 (ebk)

DOI: 10.4324/9781003392583

Typeset in Times New Roman
by Taylor & Francis Books

Contents

Illustrations

Figures

Tables

Contributors

Ibrahim Bidemi Abdullateef is a Sociologist with a keen interest in research and teaching. He is currently the lead program coordinator at Bacon Research Institute, and an international affiliate member of the American Sociological Association (ASA). Ibrahim is currently involved in a project on adolescent menstrual health in rural communities in Nigeria.

Oluwayomi Adeleke is a doctoral candidate in the Department of Sociology, University of Lagos, Nigeria. She is currently working on her final doctoral thesis. Her research interests cut across family demography, population studies and gender issues. Oluwayomi Adeleke is a laureate of the Pan African Doctoral Academy (PADA), University of Legon, Ghana.

Roberto García Alonso holds a Ph.D. from the Universidad Autónoma de Madrid, Spain. Roberto has been an associate professor at the Universidad de La Sabana in Chía, Colombia, since 2013 and is a Distinguished Fellow at the Universidad de Salamanca (Spain) (Grant Ayudas María Zambrano para la atracción de Talento Internacional funded by MCIN/AEI/10.13039/501100011033 by "European Union NextGenerationEU/PRTR". His research interests include inequality, public policies, and development in Latin America.

Ann Suwaree Ashton is an Associate Professor at the Graduate School of Tourism, National Institute of Development Administration, Bangkok, Thailand. Her expertise area is in health and wellness tourism development, the research topic includes quality of life and community-based tourism; yoga trekking for mental health well-being; spiritual retreat tourism, active aging for wellness tourist; wellness adventure tourism, sport, and leisure tourism development; slow food tourism; destination image development; and international retirement migration development.

Alessandro Bonadonna is an Associate Professor at the Department of Management, University of Turin, Italy. His research interests include local food systems and sustainable tourism. He is a research member of several projects and his teaching activities are concerned with operations and supply chain management, food quality management and environmental technologies. He is a member of the Italian Academy of Commodity Science and the Research Centre on Natural Risks in Mountain and Hilly Environments (NATRISK). He has written more than 40 scientific papers and book chapters and he is co-author of more than 10 books.

Bryan Breidenbach is a Ph.D. student and lab member of the U.S.-China Lab for Destination Development at Colorado State University in Fort Collins, Colorado, USA. Bryan's research involves a critical approach to tourism studies by focusing on the intersections of tourism destination development, conservation equity, and affordable housing strategies through means of public private partnerships. Professionally, he is interested in applying qualitative and quantitative research methods to inform housing policy within tourism dependent economies.

Carl Cater is an Associate Professor in tourism at Swansea University in Wales and visiting Professor at the University of Highlands and Islands, Scotland. His research centers on adventure tourism and ecotourism, and their contribution to development. He has written over fifty papers and book chapters, is co-author of *Marine Ecotourism: Between the Devil and the Deep Blue Sea* (2007), co-editor of the *Encyclopaedia of Sustainable Tourism* (2015) and the *Routledge Handbook of Adventure Tourism* (2024).

Tendai Chibaya lectures in the School of Hospitality, Tourism and Culture, Great Zimbabwe University, Zimbabwe. Her research interests are in tourism management, event tourism, heritage tourism, and destination marketing.

Stefano Duglio is an Associate Professor at the Department of Management, University of Turin, Italy, and Vice-President of NatRisk – Research Centre on Natural Risks in Mountain and Hilly Environments. His research interests include environmental management of the tourism industry and hospitality sector in mountain areas. Stefano is a member of the Editorial Boards of the *European Journal of Tourism Research, Tourism & Management Studies*, and *Societies* journals.

Kennedy Eborka is a Lecturer in the Department of Sociology, University of Lagos, Nigeria. His area of research and academic interests cut across population and development studies, urbanisation and community development, migration, urban poverty and conflicts in African cities. Kennedy Eborka is a Postdoctoral Fellow at the Centre for Housing and Sustainable Development (CHSD), University of Lagos, Nigeria.

Washik Muhammod Istiaq Ezaz is a Senior Lecturer in law at the Independent University, Bangladesh. His research interests revolve around the intersection of human rights and politics. Beyond academia, he embraces an active role as a human rights advocate, envisioning positive change through scholar-activism. He facilitates civil society organizations in submitting reports for the Universal Periodic Review. Washik is also a co-founder of Hearth Bangladesh and oversees the administration of the

Bangladesh Qualifying Round of the Philip C. Jessup International Law Moot Court Competition.

Rick Gage is an Assistant Professor of Natural Resources and Earth Sciences at Marshall University, USA. His research focuses on human-environment interactions, long-term benefits of adventure recreation, and the intersection of recreation and pro-environmental attitudes and behaviours. Gage is the Program Coordinator for the Natural Resources and Recreation Management degree program. He is also the founder and current Director of MAP & COMPASS, an adventure-based program designed to promote health and wellness, build social relationships, and develop leadership skills for college students.

Wendy Hillman is a Research Fellow in the Office of Indigenous Engagement, at CQUniversity, Australia, Rockhampton campus. She currently researches the amelioration of poverty in women in Nepal through transformative tourism and social entrepreneurship.

Leïla Kebir is Professor of Tourism and Territorial Economics at the Institute of Geography and Sustainability at the University of Lausanne, Switzerland. She works in the field of territorial development in the context of ecological transition. She is particularly interested in territorial resources and the way in which production and innovation are organised. She is also interested in urban and historical commons and their role in local development.

David W. Knight is co-Director of the U.S.-China Lab for Destination Development and Assistant Professor at Colorado State University in Fort Collins, USA. His research centres on tourism-linked empowerment, poverty alleviation, and sustainability to inform destination and protected area management based on resident views.

Phuong Bui L.A. is the Vice Dean of Duy Tan University's Hospitality and Tourism Institute in Vietnam. She actively engages in academic activities with international partners, focusing on tourism and social innovation, particularly in collaborations with the United States, the United Kingdom, ASEAN, and Korea. As a team leader for DTU in the British Council Vietnam's SILKEN project, she collaborates with Sapa ethnic communities to gain insights into their perspectives within the social innovation program. Her objective is to understand how local populations engage in tourism for poverty alleviation.

Ganiyu Olanrewaju Lawal is a doctoral candidate in the field of Health Studies at the University of Hull, Hull City, United Kingdom. His area of research and academic interest span the sociology of medicine and development studies. He is a member of the Black Minority Ethnic Researchers in the UK.

The-Bao Luong is a lecturer at the Faculty of Fashion and Tourism, Ho Chi Minh City University of Technology and Education, Ho Chi Minh City,

Vietnam. He is currently dedicated to advancing his career in the field of tourism, with a primary focus on gaining comprehensive knowledge of the industry and academia. His academic pursuits centre around tourism management, tourist behaviour, and cultural tourism.

Maggie C. Miller is a Professor in the School of Hospitality and Tourism Management at George Brown College, Canada. Much of her work focuses on sociocultural dimensions of sustainability and enhancing the overall quality of life for communities engaged in tourism and recreation. She is informed by relational ways of knowing and her research draws on diverse and innovative qualitive methodologies.

James Pérez-Morón is a Ph.D. candidate from Universidad de La Sabana in Chía, Colombia. James has been a professor at Universidad Tecnológica de Bolívar in Cartagena, Colombia, since 2016. His research interests include social entrepreneurship, sustainable business models, and entrepreneurship for peace.

Ravindra Nyaupane is a Senior Lecturer in Strategy and International Management at University of Roehampton, London, UK. He teaches business strategy, leadership and sustainability and his research includes doctoral studies which investigate business–society relationships and impacts. He volunteers internationally for community development and environmental projects as well as provides leadership and frontline work for charities and local causes in the UK.

Temitope J. Owolabi holds a Ph.D. from Northwest University, Mafikeng, South Africa. He is currently a Lecturer in the Department of Sociology, University of Lagos, Nigeria. His research areas include population health, industrial sociology, migration and gender studies. Temitope is a member of several international scientific communities including the International Union for the Scientific Study of Population.

Prerna Prasad is the Founder & CEO of Ecoplore, a sustainable tourism company. Ecoplore promotes and books eco-friendly hotels made of local architecture and has a minimum of 33% tree coverage inside its campus. Prerna is also the Curator of Van Bhoj: A Sustainable Living Museum in New Delhi, India. Prerna aims to invest all her income in developing a forest and returning it back to nature.

Huafang Qiao is an Associate Professor at Central China Normal University and a researcher at Wuhan Branch of China Tourism Academy. Her research interests focus on rural tourism, mountain tourism and sustainable livelihoods.

Renata Rettinger is an Associate Professor at the University of National Education Comision in Cracow, Poland. Renata's academic interests focus on issues related to determinants and consequences – as well as the process itself – of tourism development in selected regions of the world, the Caribbean in particular. Her research, in reference to sustainable growth and pro-poor tourism, concentrates mainly on enclave tourism as a prevalent form of tourism development.

Emmanuel Salim is an Assistant Professor of Geography at the University of Toulouse Jean Jaurès, France and an Associate Researcher at the University of Lausanne, Switzerland. He is interested in tourism and leisure geographies, in particular the relationship between environmental changes and the adaptation of human systems. His current research focuses on the adaptation of mountain tourism and leisure activities to environmental transformations caused by climate change. His research also explores the effects of vulnerable places on tourists' perceptions, motivations and behaviour, and discusses the emergence and dynamics of last chance and dark tourism.

Mehmet Şeremet has a PhD from the UK's Plymouth University on GIS education and employability in the UK and Turkey. His current post is that of Reader in Human Geography at Van Yuzuncu Yil University in Turkey. He leads sustainable tourism-oriented projects in different parts of Turkey, including the Black Sea Region, and the Aegean and West Mediterranean Regions. He is also the co-editor of the Turkish Geographical Review and Geographies, Planning & Tourism Studies journals. His research includes critical subjects in both human geography and tourism studies, such as critical geopolitics, border learning experience, gender and sustainability, feminist geography, and critical GIS.

Ulf Thoene holds a Ph.D. from the University of Warwick Law School, UK. Ulf has been an associate professor at the Universidad de La Sabana in Chía, Colombia, since 2013. His research has been published in the Journal of Business Ethics, the *Business and Human Rights Journal* and *Territory, Politics, Governance.* His research interests include business and human rights, business ethics, and the informal employment sector in Latin America.

Aysu Uğurlar has been an Associate Professor in the Department of City and Regional Planning at Van Yuzuncu Yil University, Turkey. Her research interests are tourism, sustainable tourism, sustainability indicators, local economic development, and urban planning.

Shuangyu Xie is a Professor and Vice Dean of the College of Urban & Environmental Sciences, Central China Normal University. She is also Vice Director of the Wuhan Branch of China Tourism Academy. Her research interests include tourism geography and regional tourism and environment.

Lina Xiong is an Associate Professor in the Department of Human Dimensions of Natural Resources at Colorado State University, USA. Her research focuses on internal branding and destination marketing with an internal stakeholder perspective. As the Director for the China Sustainable

Tourism Programs, Dr. Xiong oversees two international collaborative Masters programs – the Master of Tourism Management program – China and Master of Park and Protected Area Management program – China.

Zibanai Zhou is a PhD candidate and lecturer in the Department of Tourism, Hospitality and Leisure Sciences at Midlands State University, Zimbabwe. His main research interests span tourism development, destination marketing, event tourism, and community-based tourism.

Preface
Harnessing the World's Largest Industry to Fight Poverty and Inequality in Mountain Regions

Michal Apollo, Yana Wengel and Thomas Pogge

Tourism is one of the world's largest industries, generating an estimated 10.4% of global GDP (USD 9.2 trillion) and 10.6% of all jobs (334 million), and it was responsible for creating 1 in 4 new jobs worldwide. Moreover, international visitor spending amounted to USD 1.7 trillion in 2019 (6.8% of total exports and 27.4% of global services exports). As a result of COVID-19 and the ongoing restrictions to international mobility, the Travel and tourism sector suffered losses of almost USD 4.5 trillion, with its global contribution to GDP declining by 49.1% compared to 2019 to reach only USD 4.7 trillion in 2020; relative to a 3.7% GDP decline of the global economy. Domestic visitor spending decreased by 45%, whilst international visitor spending fell by an unprecedented 69.4%.

In the aftermath of the pandemic, however, consumer appetites for outdoor and less crowded holiday destinations have increased. Thus, mountain areas are opening up new opportunities for mountain destinations to rebuild a greener and more sustainable form of tourism and rethink their products and services. The following measures developed and described in this book will take place for this to happen. In addition, the countries sending tourists are usually countries in the Global North, and when visiting countries in the Global South, they spend their money there. Tourism is, in a way, a fair transfer of funds and thus is a powerful tool that can improve the quality of life of people in developing economies. This book provides a critical angle on mountain tourism and its sustainable use to battle poverty and inequality.

As mountain tourists, researchers, educators and poverty fighters, we genuinely care about mountain regions and the people living there, and in this edited book, we aim to address the most critical challenges. Furthermore, we aim to develop a code of conduct for tourism development of impoverished mountain regions and to suggest sustainable policy recommendations for mountain destinations, with a unique look at developing countries.

1 Mountain Tourism and Poverty Reduction

Opportunities and Challenges

Michal Apollo, Yana Wengel and Thomas Pogge

Introduction: Pro-poor concept in mountain tourism and why we use it

As the world is getting richer in aggregate, some segments of the human population have experienced substantial improvements in their standard of living, while others continue to suffer from severe poverty. In many parts of the world, poverty is even worsening as local populations are challenged by food insecurity, climate action failure, biodiversity loss, social cohesion erosion and livelihood crises, migration, digital inequality and cybersecurity failures (World Economic Forum, 2022). Disparities widen within and between countries as specific areas are disadvantaged by the environment, geographical conditions, local diseases, unenlightened cultures, or predatory rulers (Pogge, 2010). Such disadvantageous circumstances are prevalent in the world's mountain regions. Moreover, the high mountains are inhabited by some of the poorest people on Earth, who are marginalised politically and economically by national and local administrations (Messerli & Ives, 1997; Ives, 2022). Environmental constraints limit the prospects for agricultural development in many mountain areas. However, about 70% of the world's mountain population of 719 million remain rural and continue to depend on the natural resource base of land, water and forests for their livelihood (Huddleston et al., 2003). In such conditions, progress is slow, and there may be setbacks. As poverty and development are very complex phenomena, we should aim to eradicate poverty by any means, and tourism may be one of them (Apollo & Moolio, 2021; Spenceley & Meyer, 2016; Smith & Duffy, 2004; Jansky et al. 2002).

Pro-poor tourism is defined as tourism that generates net benefits for people with low incomes. Benefits may be economic, but they may also be social, environmental or cultural (Hall, 2007; Harrison, 2008; Ashley & Mitchell, 2009; Scheyvens, 2007; Spenceley & Meyer, 2016). Pro-poor tourism is not a specific product or sector of tourism but a powerful tool within the industry that aims to grow the benefits for a disadvantaged population. Tourism's role as a development tool has increased over the past four decades. Its contribution to poverty alleviation was first noted in the 1970s, but this focus was increasingly blurred in theoretical debates over 'development' in the

DOI: 10.4324/9781003392583-1

1980s and 1990s (Harrison, 2008). Thus, tourism is not a panacea to mountain poverty. Its planning and development should follow sustainable tourism tenets to enhance poverty reduction (Mutana & Mukwada, 2018). What is more, Schänzel and Apollo (2023) mentioned – in their Grand Challenge – that mountain tourism can be one of the key factors contributing to the development, prosperity, and well-being of all stakeholders.

That is why this book provides a critical angle on mountain tourism and its sustainable use to battle poverty and inequality. Mountain tourism refers to various activities in mountain settings, including mountaineering, mountain biking, winter sports activities, snowboarding, rafting, scenic viewing, wildlife viewing, ecotourism, cultural experiences and health tourism (Gill, 2022; Pomfret, 2006). Appropriately managed mountain tourism can make vital contributions to combating inequality through various tourism developments in mountain regions. Furthermore, nature-based tourism and tourism in mountain areas could provide 'healing tourism' opportunities to restore health, which in turn contributes to people's well-being; hence, in post-pandemic times, outdoor and mountain tourism might increase in popularity (Ma et al., 2021; Qiu et al., 2021; Ritpanitchajchaval, Ashton & Apollo, 2023). Overall, it is well known that mountain tourism can create sustainable development opportunities worldwide (Jansky et al. 2002; Apollo & Wengel, 2022; Ives, 2022) and that community-based tourism products provide more sustainable alternatives in destination areas than traditional mass tourism (Timothy, 2002; Mandic & Walia, 2023).

Tourism impacts mountain communities on economic, political, social, and cultural levels, often leading to changes in people's worldviews influenced by globalisation, specifically foreign cultures and lifestyles (Apollo & Andreychouk, 2022; Lu & Nepal, 2009; Nepal, 2007). Nevertheless, with a planned and systematic approach, mountaineering (subdivided into hiking, trekking, and mountain climbing) can be a critical factor in promoting overall improvement in the quality of life of the residents through initiatives in economic development and environmental conservation (Apollo & Andreychouk, 2022; Apollo & Wengel, 2022; Musa et al., 2015; Nyaupane & Timothy, 2022; Mutana & Mukwada, 2018; Wilson, & Dashper, 2023). Such promotion presupposes, however, that one can anticipate the future impact of feasible alternative paths for developing mountain tourism. Since the initiation of tourism development, various models have been created to support such predictive assessments about how such development will evolve and how it will shape stakeholder interactions (see, e.g. Ali, 2023; Apollo & Andreychouk, 2022; Mutana & Mukwada, 2018; Lu & Nepal, 2009; Nepal, 2005).

Over the last decade, international, government, non-governmental and private sector organisations have recognised that tourism can be a viable tool for addressing poverty (Hall, 2007; Pleumarom, 2012). It is generally assumed that the international tourism industry can generate economic and other benefits for poor people and communities in the context of responsible and sustainable tourism development and can thereby contribute to achieving the UN Sustainable Development Goals (SDGs) (see e.g., Saarinen, 2020; Saarinen, Rogerson & Manwa, 2013). Linking tourism to poverty reduction and the SDG agenda has led to the emergence of the 'pro-poor tourism' concept (Pleumarom, 2012). Researchers worldwide have significantly advanced this concept in recent decades (Hall, 2007; Spenceley & Meyer, 2016). However, no book has focused on mountain areas, mountain tourism, and its impact on poverty and inequality reduction. The problem is vital in that it affects 26% of the area of our planet, whose mountainous rural areas are home to 18% of the global population (authors elaboration based on Meybeck et al. 2001 & Huddleston, et al. 2003), including the 10% who are inhabiting high mountain areas (Hock & Rasul, 2019).

In the age of a neo-liberal market economy, which presents severe challenges to meeting social sustainability needs, pro-poor tourism, similar to the sustainable tourism development discourses, is shaped by a plethora of standpoints. As Esteves et al. (2012, p. 40) point out, "for the correct approach and proper understanding of the term 'social impact' we need to revisit core concepts, such as culture, community, power, human rights, gender, justice, place, resilience, and sustainable livelihoods." Here, "we the tourism people" can play a crucial role, as, due to its size, tourism impacts every sphere of our lives (Schänzel & Apollo, 2023). This book focuses on the potential pro-poor tourism has for poverty alleviation, conceiving poverty as a multi-dimensional phenomenon. Its aim is threefold: to provide a historical account of the evolution of mountain tourism as distinct from tourism in general, to offer a rich understanding of the distinctive features and contexts that make mountain tourism what it is, and to extend existing discussions of mountain tourism toward guiding future policy planning, management, and research.

Structure of the book

Comprising 17 theoretical, empirical and practical contributions, this volume deals with critical concepts, management and planning, allocating limited resources, visitors' experiences, and the future of mountain tourism. The editors have assembled authors from around the world (Fig. 1.1) to present the concept of pro-poor mountain tourism from diverse social, economic, and environmental perspectives.

The collection consists of four parts. Part I broadly outlines strategies and approaches and how these ideas can be (or are being) used to activate mountain tourism and, through it, reduce poverty. In Part II, five chapters describe the impacts and challenges of mountain tourism aimed mainly at poverty alleviation. This applies primarily to countries in the South, but countries in the North also face many challenges, as the three chapters of Part III clearly demonstrate. Part IV presents concluding remarks in two ways, i.e., through the eyes of a journalist and an academic.

Figure 1.1 Location of the study areas described in this book

Part I: Strategies for Activating Pro-Poor Mountain Tourism

The first set of chapters addresses the intervention strategies stakeholders use to harness the potential of mountain tourism as a tool for poverty reduction in developing countries. Clearly, sustainable mountain management requires a combination of conventional Western strategies with endogenous inputs that reflect the distinct cultural, social, historical, and political circumstances of each country and region (Apollo & Andreychouk, 2022; Jones, Bui & Apollo, 2021; Wengel 2020; Nepal, 2003; Nepal & Chipeniuk, 2005; Nyaupane & Timothy, 2022). Unfortunately, after decades of mountain tourism development, current management strategies have jeopardised mountain areas at risk – managers and governments need to recognise the need for change (Jones, Bui, & Apollo, 2021; Apollo & Wengel, 2022; Nyaupane & Timothy, 2022; Mandic & Walia, 2023). Management plans and guidelines developed by researchers must be considered when planning these areas' development (Apollo & Andreychouk, 2022). The chapters in this section seek answers on how to devise a development strategy that ensures that development will support the broader fight against poverty and injustice (both of which affect mountain areas disproportionally).

Miller, Cater and Nyaupane (Chapter 2) identified opportunities to harness mountain tourism for the poor through a sustainable livelihoods approach. To illustrate the issue, the authors chose the Himalayas of Nepal, which is widely known as one of the most beautiful – and one of the poorest – mountain countries in the world. For such countries, the potential benefits of tourism are substantial, but, as the authors note, the sustainable realisation of these benefits requires a careful inventory of the basic resources available to the relevant mountain communities. Miller, Cater and Nyaupane pursued this task by means of the Sustainable Livelihoods Approach (SLA), which assesses forms of natural, human, physical, financial, social, cultural, and political capital. The study indicates that tourism can support and potentially enhance these various forms of community assets, but conversely, their weaknesses will act as a barrier to development. The authors also identified suggestions for the future development of pro-poor tourism, as well as potential challenges. Their work illustrates how mountain communities can create and develop a local, sustainable economy.

In Chapter 3, Qiao, Gage, Xiong and Xie recount the genesis and evaluation of poor mountain tourism in rural areas in China. These regions still struggle with many problems and are substantially less developed than urbanised areas. The authors discuss development projects conducted by various stakeholders that reveal possible models for balancing economic prosperity and environmental sustainability. They show that pro-poor tourism in sensitive mountain environments requires coordinated planning across the private, governmental, and social spheres. Their work provides practical guidance for developing policies that support inclusive livelihoods without jeopardising sensitive cultural heritage sites and high-biodiversity habitats. Purposeful and guided management and (e.g., financial) incentives can steer development onto sustainable paths as global mountain tourism grows. The evidence that Qiao, Gage, Xiong and Xie present clearly shows that rural pro-poor tourism's unique history and continued evolution position this type of tourism as a leader in sustainable tourism development in fragile mountain areas. This approach deserves the utmost attention.

Rettinger (Chapter 4) describes tourism activation in island mountain areas. To do so, she looked at the Caribbean region, whose mountain areas are among the most isolated and, therefore, poorest parts of their respective islands/countries. It is well known that the Caribbean islands are among the mature tourist regions with a tendency for further development. However, the process of creating new or enlarging old tourism regions has thus far focused on coastal areas. Concepts of tourism development in the Caribbean islands still rely on all-inclusive tourism. Rettinger points out that the development of this type of tourism results in a strong polarisation of society and even restricts local communities from accessing the broader benefits of tourism development. The author points out that an opportunity for development (to sustain it) is activating areas beyond the coasts, namely mountainous areas. The examples of the Dominican Republic or Cuba show that tourists are eager to visit the mountainous areas of the islands. Unfortunately, the standard of living in these areas is deficient; paradoxically, these central parts of the islands are on the periphery of frequently visited coastal destinations. The mountainous regions of the Caribbean islands need tourism as a factor in socio-economic activation. As such, developing alternative destinations powered by local communities can help fight poverty in these regions.

In Chapter 5, Ashton reflects on how to use the community-based tourism development model to eradicate poverty. For this purpose, she examines how community-based tourism would affect residents' subjective, objective, and existential quality of life, using the majestic mountainous province of Loei in northeastern Thailand for illustration. Her research has shown that the catalyst for quality of life is physical, mental, emotional, social and spiritual well-being, and an effective way to create a quality of life for residents with satisfying lives in harmonious community-based tourism is to have a good quality of life and be in a community that provides a good quality of life. The theoretical conclusions of Ashton's study show a possible strategy for related stakeholders to apply and develop to suit their businesses. However, the author points out that the strategy will be effective with government support, such as developing tourism infrastructure, including environmental sustainability, building attractive communities, high-standard tourism facilities, and good quality accommodation or homestay accommodation.

Pérez-Morón, Thoene and Alonso (Chapter 6) address the dark side of tourism in Colombia's mountain areas. Their essay is an essential step toward illustrating the potential for tourism to be used as a development path for post-conflict areas. In many parts of the world, mountain territories have been used as sites of armed conflict and refuge for illegal actors. These areas, often notorious, have become isolated. However, the authors argue that mountain tourism appears to be an opportunity for local development that rebuilds mountain people's and former combatants' lives. This chapter is one of the first attempts to shed new light on dark tourism as a way of reintegrating former combatants and victims of Colombia's armed conflict as an alternative for mountain dwellers in a developing post-conflict country. Pérez-Morón, Thoene and Alonso's goal is to show how tourism can be used as a tool to reintegrate former combatants and victims of armed conflict in mountainous regions. Their work also highlights ecotourism's potential to create economic, environmental, and social opportunities for mountain residents.

In Chapter 7, Hillman addresses the situation of women in mountain tourism. The author analyses the links between pro-poor mountain tourism and women's employment in Nepal, showing that both women and men are exposed to inequalities resulting from the deprivations and vulnerabilities the mountain environment gives rise to. Hillman also points out, however, that women also endure special societal and spiritual prejudices about their capabilities, which impede their rightful involvement in community, civil, and commercial activities and denigrate the assets, encounters, and inspiration of half the world's inhabitants. The chapter discusses the benefits and pitfalls of various types of tourism related to the pro-poor sphere. Hillman highlights the critical and multifaceted roles played by government, private sector players, NGOs, community organisations, and the people within pro-poor mountain tourism. The chapter concludes with suggestions on how to guide the development of pro-poor mountain tourism and the inclusion of impoverished people, especially women. This critical and essential chapter concludes the first part of the book.

PART II: Impacts and Challenges of Taking a Pro-Poor Approach to Mountain Tourism

The second set of chapters deals with mountain tourism's impact and challenges and focuses on poverty reduction in developing countries. Overall, any tourism activity in a fragile mountain environment causes certain impacts and challenges in every sphere of the mountains, both environmental (Apollo, 2021; Jones, Bui, & Apollo, 2021) and socio-economic (Apollo & Wengel, 2022). However, in contrast to the aforementioned items in the literature, launching, developing, or directing mountain tourism into poverty alleviation tracks can bring several additional challenges and impacts (Apollo & Andreychouk, 2022; Mutana & Mukwada, 2018). The authors of the following chapters attempt to fill this knowledge gap.

In Chapter 8, Uğurlar and Şeremet focus on issues of socio-cultural integration and its impact on the development of pro-poor mountain tourism. To this end, the authors make a historical assessment of the Kyrgyz community's tourism potential in the context of the mountain environment. The chapter shows that the government initially supported the mountain community through social policies (housing, jobs, etc.). However, as Uğurlar and Şeremet point out, the second and third generations deserve the better living conditions and opportunities that big cities offer. This study shows that skills and knowledge are among the most important assets that help mountain communities create a cultural-based tourism facility. It is worth mentioning here that while many studies in the literature focus on the role of pro-poor tourism in alleviating poverty and inequality, this study provides critical insight into the socio-cultural integration of international communities in Turkey with pro-poor tourism. Such integration also facilitates their communication with both local Turkish and Kurdish communities. The chapter provides a set of practical and theoretical implications for improvement.

In Chapter 9, Eborka, Lawal, Abdullateef, Owolabi and Adeleke discuss the prospects and challenges of developing tourism in Nigeria's mountainous areas. The authors explain the enormous potential that Nigeria's mountain tourism holds while showing that the country has not yet fully realised this potential. The authors pinpoint several factors that hinder mountain area development. Some of these include the neglect and exclusion of mountain communities in the management of mountain tourism, widespread institutional corruption, lack of security and, unfortunately, lack of government will. The chapter notes that while mountain tourism appears to be positively impacting the mountain communities of this African country, there is still much work to be done to properly realise the potential of mountain tourism. This critically important article shows that there can be massive potential in countries that, like Nigeria, are not known for mountain tourism. Moreover, the authors urge the Nigerian government to learn from the examples of countries that have effective mechanisms for managing the tourism sector in order to develop the country and to improve the quality of life of its communities.

Chibaya and Zhou reach similar conclusions in Chapter 10. The authors show how to use mountain tourism to reduce poverty in Zimbabwe's Eastern Highlands. Although when we think of mountains in Africa, we usually think of Tanzania's Kilimanjaro, the continent is full of scenic mountain areas that can be used to combat poverty in isolated mountainous areas. Unfortunately, as Chibaya and Zhou point out, pro-poor mountain tourism in developed economies has received much attention in the pro-poor tourism literature, while the literature on developing countries is still sparse. As such, this chapter (like Chapter 9) takes a visionary approach and shows how pro-poor mountain tourism can be optimised to address poverty issues. It also uncovers the challenges inhibiting African mountain communities from taking full advantage of pro-poor mountain tourism enterprises. The analysis results showed that only inclusive mountain tourism can significantly contribute to the fight against poverty. However, lack of funding and know-how, climate change, biodiversity destruction, and excessive leakage have hindered the success of mountain tourism. It is recommended that mountain community residents be empowered by being given the financial resources to play a leading role in mountain tourism enterprises and capacity-building programs.

Bui L.A. and Luong (Chapter 11) address ethnic tourism and how it can be used and guided to alleviate poverty and inequality. To illustrate the problem, the authors focused on the Sapa district, located in the highlands of northwestern Vietnam, which is famous for its picturesque landscapes and cultural diversity. Focusing on the use of tourism as a strategy to alleviate poverty and improve the lives of the local ethnic group, the chapter identifies challenges and opportunities. The authors show that tourism has positively impacted the lives of the ethnic population, improving income and employment opportunities. However, they also demonstrate challenges to ensuring sustainable tourism development in the Sapa district, including economic, social, environmental, and cultural issues. Bui L.A. and Luong recommend the use of ethnic tourism, which, with the many opportunities and challenges it offers, holds great potential for poverty alleviation as well as for preserving the region's heritage and the sustainability of its natural environment.

Chapter 12 also shows the potential of ethnic tourism. Underscoring the importance of human rights in a pro-poor approach to mountain tourism, its author, Ezaz, based his discussion on the Chittagong Hill Tracts tribal area of Bangladesh, which struggles with poverty and marginalisation. Given the location and remoteness of the area and its associated lack of industry and infrastructure, its population depends on subsistence farming. However, its natural beauty makes it possible for tourism to become a significant contributor to development and poverty alleviation. As the author points out, existing development models, without consideration of human rights issues, have proven inadequate in the fight against poverty. Ezaz also sketches how a human rights-based approach can effectively deal with systemic barriers and promote sustainable development by prioritising the rights and needs of marginalised tribal communities. By proposing a human rights-based framework that emphasises participation, equality, and environmental protection as the best path to sustainable development, the author advocates empowering marginalised tribal communities while respecting their cultural heritage. Respect for human rights can overcome some barriers to alleviating poverty.

Part III: Lessons from Advanced Economies

Nowadays, mountains attract people not only because of their natural resources but also because of the unique aesthetic and recreational values associated with the landscapes (Apollo & Andreychouk, 2022; Mutana & Mukwada, 2018). Therefore, tourism has become a significant form of human economic activity in the mountains, especially in economically developed countries (Price, 1992). In general, the impact of and the issues raised by mountain tourism development are not the same in developing and developed countries (Nepal, 2002; Lew & Han, 2015). Developing countries are looking for a way to take advantage of the lessons of developed countries toward overcoming widespread poverty as well as economic and cultural marginalisation (Part I and II). It should be noted, however, that such marginalisation has not been overcome everywhere in developed countries and that prosperity, once achieved, may not last. For example, the European Alps experienced a loss of cultural attractiveness when the mountain region became overbuilt and overdeveloped between the 1960s and 1980s (Price et al., 1999). Another extremely important element may be climate change, which dramatically affects the reality of mountain tourism (Ives, 2005; François et al., 2023; Salim, 2023; Steiger et al., 2023). The following three chapters address these issues, showing the reader that pro-poor mountain tourism strategies are also needed in some developed countries.

In Chapter 13, Breidenbach and Knight discuss the role of tourism in the fight against poverty in developed countries. In their work, the authors describe Mountain West communities in the US experiencing many new problems associated with a transforming economy, from extractive to those dependent on tourism. To understand the many dimensions of sustainable tourism development in such communities, Breidenbach and Knight address the role of tourism in poverty alleviation and the potential for using sustainable tourism to deal with growing wealth inequality and lack of housing opportunities. Determining how sustainability, resilience, motivations and conflicts can be synergised requires extensive temporal and spatial fields to recognise synergies that improve the application of sustainable tourism development frameworks. There is no panacea for solving all issues related to sustainable tourism development, but further research offers tools that can be used to promote a better future. The chapter concludes with an overview of management implications and recommendations for future research, offering key insights for combating tourism-related poverty while optimising outcomes for the most vulnerable.

In Chapter 14, Bonadonna and Duglio focus on local food markets as elements supporting sustainable mountain tourism in peripheral areas. They focus on the northeast part of the Piedmont region (Italy), which lacks iconic mountains and major ski resorts. Nevertheless, this region has seen increased nature and outdoor tourism in recent decades. The authors explore the potential for additional regional promotion by using a local product – Bettelmatt cheese, produced according to a traditional recipe – which can be an essential part of the tourist offer. The chapter thus aims to highlight the socio-economic opportunities associated with the Bettelmatt brand by focusing on how it can generate income and stimulate sustainable tourism. Bonadonna and Duglio thereby illustrate and provide a valuable example of how the development of a traditional production system (in this case, food) in a local area offers numerous economic opportunities but also requires careful management of relationships, environmental and social implications, and available resources. Area managers must be prepared to meet these challenges by adopting strategies and making decisions that ensure the local community's social, economic, and environmental success.

The final chapter (Chapter 15) of this section deals with tourism in the era of climate change, one of the most critical elements that, in both developing and developed countries, will shape the reality of mountain areas. Authors Salim and Kebir approach this problem by describing a regional perspective on the transformation of glacier tourism in the era of climate change. The authors provide various examples from around the world of how these climate changes affect the economy of mountain areas, mainly focusing on four glaciers in the Alps (France, Switzerland and Austria). Their historical analysis illustrated the evolution of the locals' approach to glacier tourism, which has led to the emergence of international resorts or tourist destinations. As the authors' research and the works they cite indicate, current climate change, which is causing a drastic shrinking of glaciers worldwide, may change their position. If this thesis comes from the developed world, and the situation is so noticeable and can have such severe consequences for the prosperous North, one can only guess how serious a challenge the South will face. Salim and Kebir conclude their chapter by stating that the effects of climate change may have severe consequences for glacier tourism and mountain tourism in general.

PART IV: Concluding Remarks

In the last part, we revisit the book's key points. The reader, already having an overview of the issues related to pro-poor mountain tourism, will be familiar with the summary, which will outline the main aspects and identify future testing grounds. The authors reflect on the complexities and shortcomings of the current pro-poor mountain tourism policy, as well as development in the broadest sense. After all, it is no mystery that most people support development and often use the word as a slogan for what they simply consider good (Payne & Phillips, 2010). But, what is it? Arndt (1989) noted that when educated citizens from developed and developing countries are asked what they understand to be the desired goal of the development process, they give very different answers. This is mainly because all parties involved (in this case, hosts, tourists and researchers) have different points of view regarding tourism and the development and changes in destination areas related to the tourism industry. This is also what emerges from this book. Arguably, nature-based tourism in developing countries is economically driven, based mainly on leisure, and characterised by resorts that use nature, leading to overdevelopment (Dolezal et al., 2020). People can contribute to the "enhancement" of nature for their consumption and convenience towards enlightened mass tourism in the mountains (Jones et al., 2021; Weaver et al., 2020), but appropriate and sustainable management is required in all cases (Apollo et al., 2020; Apollo & Wengel, 2022). Therefore, reinforcing the vision and guidance provided by the authors in Parts I-III, Part IV adds a reflection on the future of mountain regions and their sustainable development.

In Chapter 16, Prasad evaluates pro-poor mountain tourism and does so with a journalist's eye. In doing so, he generalises his knowledge by demonstrating the most important courses of action. On the other hand, in Chapter 17, the volume editors Apollo, Wengel and Pogge provide a scientific summary along with their thoughts on future prospects for poverty reduction strategies using mountain tourism.

A Preview of the Approach

This book aims to reflect on the potential of mountain tourism for pro-poverty action, to discuss the impacts of mountain tourism from an environmental, socio-cultural, and personal perspective, and to identify current management policies. It links the past and future of mountain tourism, and it provides policy suggestions for sustainable management in fragile mountain regions.

Based on the chapters' focus, editors and authors seek to offer conclusions that address four competing and sometimes conflicting moral considerations:

- the economic and other interests of the local mountain dwellers,
- the economic and other interests of potential visitors,
- the economic and other interests of the rest of the country and its governing bodies,
- the global interest in preserving the regional environment.

Thereby, through a series of cases and research from experts in the field, this book assesses the effectiveness of pro-poor mountain tourism as a development strategy. It tackles the issue of whether mountain areas of developing and developed countries benefit from tourism's potential role in poverty and inequality reduction. After all, tourism is not just about fulfilling the selfish satisfaction of those who visit and pay (see e.g., Smith & Duffy, 2004). The ethical use of this potential tool of tourism should play an important role in combating injustice and poverty. Thus, can mountain tourism 'afford' morality?

References

Ali, A. (2023). Estimating the recreational value of mountain tourism to shape sustainable development in Gilgit-Baltistan, *Pakistan. Journal of Cleaner Production*, 426, 138990. https://doi.org/10.1016/j.jclepro.2023.138990.

Apollo, M., & Andreychouk, V. (2022). *Mountaineering Adventure Tourism and Local Communities: Social, Environmental and Economic Interactions.* Edward Elgar Publishing. https://doi.org/10.4337/9781802209389.

Apollo, M., & Moolio, P. (Eds.) (2021). *Poverty and Development: Problems and Prospects.* Channel View Publications. https://doi.org/10.21832/9781845418489.

Apollo, M., & Wengel, Y. (2022). *Mountaineering tourism: A critical perspective.* Routledge. https://doi.org/10.4324/9781003095323.

Apollo, M., Andreychouk, V., Moolio, P., Wengel, Y., & Myga-Piątek, U. (2020). Does the altitude of habitat influence residents' attitudes to guests? A new dimension in the residents' attitudes to tourism. *Journal of Outdoor Recreation and Tourism*, 31, 100312https://doi.org/10.1016/j.jort.2020.100312.

Arndt, H. W. (1989). *Economic development: The history of an idea.* University of Chicago Press.

Ashley, C., & Mitchell, J. (2009). *Tourism and Poverty Reduction: Pathways to Prosperity* (1st ed.). Routledge.

Esteves, A. M., Franks, D., and Vanclay, F. (2012). Social impact assessment: the state of the art. *Impact Assess. Proj. Apprais* 30, 34–42. https://doi.org/10.1080/14615517.2012.660356.

François, H., Samacoïts, R., Bird, D. N., Köberl, J., Prettenthaler, F., & Morin, S. (2023). Climate change exacerbates snow-water-energy challenges for European ski tourism. *Nature Climate Change*, 13(9), 935–942. https://doi.org/10.1038/s41558-023-01759-5

Gill, A. M. (2022). Mountain Tourism. In D. Buhalis (Ed.), *Encyclopedia Of Tourism Management And Marketing.* Edward Elgar Publishing. https://doi.org/10.4337/9781800377486.mountain.tourism.

Hall, C. M. (2007). *Pro-Poor Tourism: Who Benefits? Perspectives on Tourism and Poverty Reduction.* Channel View Publications.

Harrison, D. (2008). Pro-poor Tourism: A Critique. *Third World Quarterly.* 29(5), 851–868. https://www.jstor.org/stable/20455079.

Hock, R., & Rasul, G. (2019). High Mountain Areas. In H.-O. Pörtner, D. C. Roberts, V. Masson-Delmotte, P. Zhai, M. Tignor, E. Poloczanska, K. Mintenbeck, A. Alegría, M. Nicolai, A. Okem, J. Petzold, B. Rama, & N. M. Weyer (Eds.), *Special Report on the Ocean and Cryosphere in a Changing Climate* (pp. 133–151). IPCC.

Huddleston, B., Ataman, E., De Salvo, P., Zanetti, M., Bloise, M., Bel, J., … & Fed'Ostiani, L. (2003). *Towards a GIS-based analysis of mountain environments and populations.* Environment and Natural Resources working paper, 10. https://www.fao.org/3/y4558e/y4558e.pdf.

Ives, J. D. (2005). Personal Views: Global Warming: A Threat to Mount Everest?. *Mountain Research and Development*, 25(4), 391–394. https://www.jstor.org/stable/3674451.

Ives, J. D. (2022). *Sustainable mountain development: Getting the facts right.* Springer Nature. https://doi.org/10.1007/978-3-030-96029-2.

Jansky, L., Ives, J. D., Furuyashiki, K., & Watanabe, T. (2002). Global mountain research for sustainable development. *Global Environmental Change*, 12(3), 231–239. https://doi.org/10.1016/S0959-3780(02)00015-8.

Jones, T., Bui, H., & Apollo, M. (2021). *Nature-based Tourism in Asia's Mountainous Protected Areas: A Trans-Regional Review of Peaks and Parks.* Springer International Publishing. https://doi.org/10.1007/978-3-030-76833-1.

Lew, A. A., & Han, G. (2015). A world geography of mountain trekking.In G. Musa, J. Higham, & A. Thompson-Carr (Eds.), *Mountaineering Tourism* (19–39). Routledge. https://doi.org/10.4324/9781315769202.

Lu, J., & Nepal, S. K. (2009). Sustainable tourism research: An analysis of papers published in the Journal of Sustainable Tourism. *Journal of Sustainable Tourism*, 17(1), 5–16. https://doi.org/10.1080/09669580802582480.

Ma, S., Zhao, X., Gong, Y., & Wengel, Y. (2021). Proposing "healing tourism" as a post-COVID-19 tourism product. *Anatolia*, 32(1), 136–139. https://doi.org/10.1080/13032917.2020.1808490.

Mandić, A., & Walia, S. K. (Eds.). (2023). *The Routledge Handbook of Nature Based Tourism Development*. Routledge. https://doi.org/10.4324/9781003230748.

Messerli, B., & Ives, J. D. (1997). *Mountains of the World: A Global Priority.* Parthenon Publishing Group. https://doi.org/10.2307/3060455.

Meybeck, M., Green, P., & Vorosmarty, C. (2001). A new typology for mountains and other relief classes. *Mountain Research & Development*, 21 (1), 34–45. https://doi.org/10.1659/0276-4741(2001)021[0034:ANTFMA]2.0.CO;2.

Musa, G., Higham, J., & Thompson-Carr, A. (2015). Mountaineering Tourism. In G. Musa, J. Higham, & A. Thompson-Carr (Eds.), *Mountaineering Tour.* Routledge. https://doi.org/10.4324/9781315769202.

Mutana, S., & Mukwada, G. (2018). Mountain-route tourism and sustainability. A discourse analysis of literature and possible future research. *Journal of outdoor recreation and tourism*, 24, 59–65. https://doi.org/10.1016/j.jort.2018.08.003.

Nepal, S. (2005). Tourism and remote mountain settlements: Spatial and temporal development of tourist infrastructure in the Mt Everest region, Nepal. *Tourism Geographies*, 7(2), 205–227. https://doi.org/10.1080/14616680500072471.

Nepal, S. K. (2002). Mountain ecotourism and sustainable development. *Mountain research and development*, 22(2), 104–109. https://doi.org/10.1659/0276-4741(2002)022[0104:MEASD]2.0.CO;2.

Nepal, S. K. (2007). Tourism and rural settlements Nepal's Annapurna region. *Annals of Tourism Research*, 34(4), 855–875. https://doi.org/10.1016/j.annals.2007.03.012.

Nyaupane, G. P., & Timothy, D. J. (Eds.) (2022). *Tourism and development in the Himalaya*. Routledge. https://doi.org/10.4324/9781003030126.

Payne, A., & Phillips, N. (2010). *Development*. Polity Press.

Pleumarom, A. (2012). *The Politics of Tourism, Poverty Reduction and Sustainable Development*. Third World Network.

Pogge, T. (2010). *Politics as Usual. What Lies Behind the Pro-Poor Rhetoric*. Polity Press.

Pomfret, G. (2006). Mountaineering adventure tourists: a conceptual framework for research. *Tourism management*, 27(1), 113–123. https://doi.org/10.1016/j.tourman.2004.08.003.

Price, M. F. (1992). Patterns of the development of tourism in mountain environments. *GeoJournal*, 27, 87–96.

Price, M., Wachs, T., & Byers, E. (1999). *Mountains of the world: Tourism and sustainable mountain development*. Institute of Geography, University of Berne.

Qiu, M., Sha, J., & Scott, N. (2021). Restoration of visitors through nature-based tourism: A systematic review, conceptual framework, and future research directions.

In *International Journal of Environmental Research and Public Health* (Vol. 18, Issue 5, pp. 1–19). https://doi.org/10.3390/ijerph18052299.

Ritpanitchajchaval, N., Ashton, A. S., & Apollo, M. (2023). Eudaimonic well-being development: Motives driving mountain-based adventure tourism. *Journal of Outdoor Recreation and Tourism*, 42, 100607. https://doi.org/10.1016/j.jort.2023.100607.

Saarinen, J. (Ed.). (2020). *Tourism and sustainable development goals: Research on sustainable tourism geographies.* Routledge. https://doi.org/10.4324/9780429324253.

Saarinen, J., Rogerson, C. M., & Manwa, H. (Eds.). (2013). *Tourism and the millennium development goals: Tourism, local communities and development.* Routledge. https://doi.org/10.4324/9781315868530.

Salim, E. (2023). Glacier tourism without ice: Envisioning future adaptations in a melting world. *Frontiers in Human Dynamics*, 5, 1137551. https://doi.org/10.3389/fhumd.2023.1137551.

Schänzel, H. & Apollo, M. (2023). Specialty grand challenge: Wherever tourism arrives, society notes it first! *Frontiers Sustainable Tourism* 2, 1341765. https://doi.org/10.3389/frsut.2023.1341765.

Scheyvens, R. (2007). Exploring the tourism-poverty nexus. *Current Issues in Tourism* 10(2/3), 231–254. https://doi.org/10.2167/cit318.0.

Smith, M. & Duffy, R. (2004). *The ethics of tourism development.* Routledge. https://doi.org/10.4324/9780203634325.

Spenceley, A., & Meyer, D. (Eds.). (2016). *Tourism and poverty reduction: Principles and impacts in developing countries.* Routledge. https://doi.org/10.4324/9781315677057.

Steiger, R., Knowles, N., Pöll, K., & Rutty, M. (2022). Impacts of climate change on mountain tourism: A review. *Journal of Sustainable Tourism*, 1–34. https://doi.org/10.1080/09669582.2022.2112204.

Timothy, D. J. (2002). Tourism and community development issues. In R. Sharpley & D. J. Telfer (Eds.), *Tourism and development: Concepts and issues* (pp. 149–164). Channel View Publications.

Weaver, D., Tang, C., & Zhao, Y. (2020). Facilitating sustainable tourism by endogenisation: China as exemplar. *Annals of Tourism Research*, 81. https://doi.org/10.1016/j.annals.2020.102890.

Wilson, J., & Dashper, K. (2023). In the shadow of the mountain: The crisis of precarious livelihoods in high altitude mountaineering tourism. *Journal of Sustainable Tourism*, 31(10), 2270–2290. https://doi.org/10.1080/09669582.2022.2108038.

Part I

Strategies for Activating Pro-Poor Mountain Tourism

2 Identifying Pro-Poor Mountain Tourism Opportunities Through a Sustainable Livelihood Approach

A Case Study of Dhorpatan Valley, Nepal

Maggie Miller, Carl Cater and Ravindra Nyaupane

Introduction

Nepal, a landlocked country situated between China and India, has a population of over 30 million (World Bank, 2023). Understood as a low-income economy, this nation is ranked 143rd of 191 nations for overall level of socio-economic development on The United Nations Human Development Index (HDI) (United Nations Development Programme, 2023). While its global position has improved slightly since 2020, 28 percent of Nepal's rural dwellers, who depend primarily on agriculture, still remain multidimensionally poor, compared with 12.3 percent of the population in urban areas (Government of Nepal, 2021). Indeed rapid increases in workforce out-migration are the corollary of these impoverished conditions, as well as the nation's poor resource base and low industrial and services output (Bhattarai et al., 2005; Nepal, 2003, 2005). This workforce drain has impacted rural hill regions of Nepal in particular, changing the demographics and socio-economic fabrics of their communities (Nyaupane et al., 2019).

Because of their geography and often their remoteness, development in mountain regions has historically lagged behind those of lowland areas. Yet research suggests that the biophysical base of a region often dictates the kind of developments that are available, while these places and their economic histories are considered among national, continental, and global economies (Mair, 2006). Situated in the heart of the Himalayas, eight of the fourteen world-famous mountain peaks over 8000 meters tall are located in Nepal. By no surprise, mountain tourism was embraced as the "saviour" – a core development strategy – since Nepal opened its borders to foreigners in 1951, and its rapid growth has transformed the nation in unprecedented ways (Nepal, 2005). These stories of development align with those of other developing countries as international tourism is said to hold much promise. Mair (2006) recognises this as the "conundrum of tourism" and critiques these development strategies for their sustainability, where tourism is often positioned as a "way for communities to survive when there are no other opportunities" (p. 31).

DOI: 10.4324/9781003392583-3

Though mountain tourism has the potential to create prosperity in mountainous regions, Apollo and Wengel (2022) maintain that sustainable development depends on conservation of resources and participation of local communities. Yet, much of the literature around pro-poor tourism notes that whilst the poor may own the natural and cultural assets needed for tourism they may not always be recognised, and communities may not have the skills to exploit them (Telfer & Sharpley, 2016). Therefore, we use this chapter to identify baseline assets available to the rural communities and villages located in the high-altitude Dhorpatan valley of western Nepal. To appraise such assets, we employ a *Sustainable Livelihoods Approach* (SLA) which evaluates forms of natural, human, physical, financial, social, cultural and political capital. Our objective here is to examine the SLA in the Dhorpatan region in order to support the planning and development of a more local, sustainable economy. Prior to this examination we outline the SLA framework, and its application and utility in other regional and tourism studies. After appraising these resources, we extend suggestions for future pro-poor tourism development in Dhorpatan and identify potential challenges for consideration.

Pro-Poor Tourism and The Sustainable Livelihood Approach

A pro-poor approach towards tourism development endeavours to contribute to poverty reduction and enhance the capacity of the poor to participate in tourism product development and decision making (Ashley & Roe, 2002; Scheyvens, 2007). Such approaches recognise that whilst communities may be the hosts of tourism, benefits rarely accrue to the poorest members. There is therefore a "need to develop policies that spread the benefits of tourism development to the poorest members of destination societies who are unable to establish a formal position in the local tourism system" (Telfer & Sharpley, 2016, p. 66). Paradoxically tourism is often proposed as a development option as undeveloped communities often have the assets (usually cultural or natural) that can support tourism. Identifying these assets is therefore the first step in empowering communities and alleviating poverty; one such tool to do so is the Sustainable Livelihoods Approach (SLA).

The SLA is a pro-poor approach to appraising community assets and their development for vulnerable societies (Cater et al., 2015). It emphasises that "the root of all human development and economic growth is livelihoods – not jobs per se, but the wide, infinitely diverse range of activities people engage in to make their living" (Tao & Wall, 2009, p. 143). Building upon the original concept of "sustainable livelihood" (SL) as described by Scoones (1998), the SLA is a dynamic lens that highlights "the changes and adjustments people make in their livelihood systems in order to cope under difficult circumstances" (Helmore & Singh, 2001, p. 3). This approach also stresses that using internal capacities and knowledge systems help the development of rural communities (Lee, 2008).

The SLA is a useful framework to test changes wrought by tourism on people's asset base. The approach, which aims to reduce poverty in developing countries, makes a systematic evaluation of the variety of assets in a community that might be used and augmented by tourism development (Cater & Cater, 2007). Ashley's (2000) study, for example, offers two perspectives on rural communities using tourism for their livelihoods. Previous tourism developments have tended to focus primarily on the contribution to enhancing local employment and income benefits, often to the exclusion of the major social and environmental changes in livelihoods (Wang et al., 2016). Under the SLA, however, the focus is on how to enhance development and the contribution tourism can make, while also reflecting on the broad changes to people's livelihoods consequent to development decisions. The approach is a multi-level one and regarded as dynamic for households, communities, regions, or nations (Cater & Cater, 2007).

Initial work on the SLA identified five types of assets or capital central to people's livelihoods including: natural, human, physical, financial and social. However, in a global tourism context, where local traditions form both the uniqueness and identity of communities, Cater and Cater (2007) identified a further asset: cultural capital. This type of capital consists of community heritage, customs, and traditions. Wang and colleagues (2010) added political capital to the sustainable livelihoods approach to further explain governance and administrative structures in community-based tourism development. Political capital identifies elements of power and benefit sharing as issues of concern for community stakeholders, which has potential to affect all the other forms of capital in the sustainable livelihood approach.

Forms of Capital

The seven forms of community assets under the SLA can therefore be identified as:

- a natural capital: natural resource stocks upon which people draw for livelihoods;
- b human capital: skills, knowledge, ability to labour, and good health that are required to pursue different livelihood strategies;
- c physical capital: basic enabling infrastructure, such as transport, shelter, water, energy and communications;
- d financial capital: financial resources available to people such as savings, credit, and remittances;
- e social capital: social resources such as networks, membership of groups, and relationships of trust upon which people draw in pursuit of their livelihoods;
- f cultural capital: heritage, customs, festivals, and traditions of the community; and
- g political capital: governance and administrative structures.

These seven forms of capital are highly interdependent, as depicted in Figure 2.1. Pictured are the building blocks of the SLA whereby the financial, social, and cultural capital constitute the community foundation. Physical infrastructure and human skills require the underpinning of the financial, social, and cultural capital. The top of the heptagon, the natural capital, central to mountain tourism, relies on sustained and sustainable support from all of these forms of capital. If the natural capital itself is eroded, perhaps by external influences, then the viability of other forms of capital in the community will also be in doubt.

Dhorpatan Area

The Dhorpatan area is set in the foothills of the Himalayan mountains in western Nepal. The Dhorpatan municipality is centred on the town of Burtibang and has nine villages (formally referred to as wards). Dhorpatan is a highly rural, hilly area and the average altitude of farming land is between 2000 to 3000 meters. The economy is heavily reliant on subsistence farming supported by limited trade, remittance from migrant workers, and foreign aid. Dhorpatan is also overlapped by the Dhorpatan Hunting Reserve, a protected area administered by the National Trust for Nature Conservation and the only hunting reserve in Nepal. The altitude of the reserve ranges from 2850 to 5500 meters above sea level (Panthi et al., 2012). Permit fees to enter the hunting reserve are the main indicator of tourism numbers which are low to date, and indeed have declined from the mid-1990s when an average of 200 tourists visited the reserve, with no recorded visitors at all from 2003–2007, 2010, 2011 and 2015. (Nepal Ministry of Culture, Tourism, & Civil Aviation, 2020). Low numbers could reflect the remoteness of the mountainous

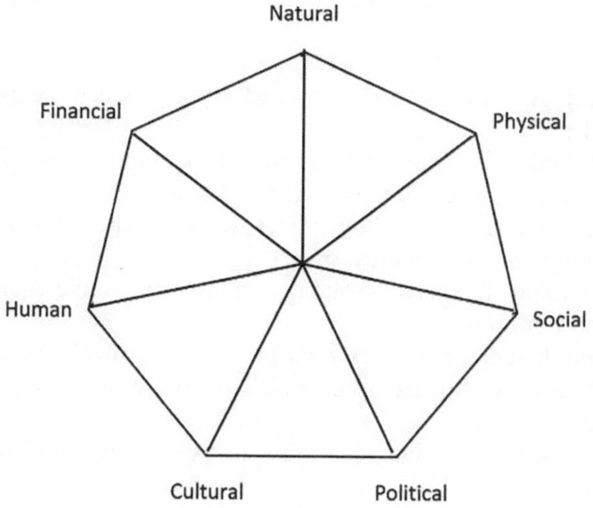

Figure 2.1 The elements of the sustainable livelihoods approach (Wang, et al., 2016)

location, where access is much more complex than for other tourism destinations (Apollo, 2017; Apollo & Wengel, 2022). Of the tourists that Dhorpatan does see, they tend to visit in the two popular tourist seasons of April/May and October/November when the weather is most suitable.

Regional multi-stakeholder tourism and livelihoods initiatives are supported by organisations like the United Network, an NGO who formed the Dhorpatan Cluster to establish a regional identity and local approach to conservation and livelihood development (Nyaupane et al., 2019). Integral to the Dhorpatan Cluster are the communities that reside in the Dhorpatan valley – the area with which this chapter is primarily concerned. The Dhorpatan valley is a distinct high-altitude valley, defined by the north-flowing Uttarganga river (see Figure 2.2), located some 100km from Baglung and 15km from Burtibang., and has an approximate population of 7000 people. It consists of the following main villages:

- Navi Village (Approx. population 2500; 450 homes)
- Pathathar Village (Approx. population 500; 75 homes, primarily Tibetan refugees)
- Chhyantung Village (Approx. population 200; 45 homes)
- Bhuji and Vani Villages (Approx. population 175; 37 homes)

Figure 2.2 Uttarganga river and Dhorpatan Valley. Photograph by Carl Cater.

Current livelihood activities in these villages are related to agriculture and animal husbandry with a few home stays and small hotel businesses. Additionally, the Uttarganga river supports small hydropower sites for some villages, and has been recognised as a potential resource for future hydropower development (e.g., National Uttarganga Hydropower project), and an attraction to be developed for tourism activities. This chapter will now explore the community assets and capital in detail.

Dhopatan Valley Livelihood Assets

The SLA framework was used to identify baseline assets available to villages within the Dhorpatan valley. To assess and evaluate forms of natural, human, physical, financial, social, cultural and political capital, data and insights were gathered from secondary resources and first-hand observations. One of the authors has visited the area numerous times and is very familiar with the region and the other two authors conducted a short scoping trip in April 2019 as part of a student field trip. Along with field observations and conducting interviews with community members, we also held a multi-stakeholder forum to identify challenges in the area. It is important to note that existing research and literature around the valley is scant, making it slightly more challenging to compose a comprehensive picture of its livelihood assets. Our fieldwork was also exploratory at this stage, and we do not present individual community members views here. This, of course, highlights opportunities for further research in this region.

Natural capital

Natural Capital is the natural resource stocks upon which people draw for their livelihoods and is the cornerstone of the SLA approach (Wang et al., 2016). To a large extent, Dhorpatan's natural capital is governed by its geography as a high-altitude mountain environment. The climate is dictated by this geography, the monsoon season in this area occurs between June and September, but rains may be lighter than areas further south of the mountains, with annual precipitation of <1000 mm (Kandel, 2000). Higher elevations are subject to strong winds throughout the year, and snow usually lasts, even at lower elevations, until early April.

Balanced weather conditions in the region allow for specialised agriculture, livestock, and tourism activities. Because of its altitude, current and potential crops differ from those in lower regions and include more temperate species such as potatoes, kodo millet, cardamom, apples, and oranges (Nyaupane et al., 2019). Alpine meadows above the treeline are locally known as "patans" and are important for both indigenous animals such as the blue sheep but also for livestock grazing. There are concerns with levels of transhumance activity as some livestock grazing originates from many days away.

The Dhorpatan Hunting Reserve is characterised by alpine, sub-alpine, and high-temperate vegetation. The reserve supports red panda, blue sheep, snow leopard, Himalayan tahr, Himalayan black bear, barking deer, wild boar, rhesus macaque and wolf (Panthi et al., 2012). Yarsagumba (Ophiocordyceps sinensis), a parasitic fungus that develops on dead caterpillars is also present in the region and an important export resource. Endemic to the Himalayas and Tibetan Plateau, it is one of the most valuable medicinal mushrooms in the world (earning up to US$100,000/kg). Yarsagumba is collected in late spring to early summer after snow starts thawing in the alpine and subalpine pastures. There are sustainability concerns around harvesting this fungus, with declining yields observed in recent years, leading to an informal licensing and revenue generation system (Thapa et al., 2014).

Human capital

Human capital is based on skills, knowledge, ability to labour, and good health that are important to be able to pursue different livelihood strategies (Cater et al., 2015). Subsistence agriculture is the mainstay of the Dhorpatan valley economy, and most of the local community skills are related to farming (Figure 2.3). Education provision is poor within the valley, presently there is only one primary school (with limited subjects taught) and one private boarding school, run by the Tibetan refugee community in the Pathathar village.

Figure 2.3 Farming in Dhorpatan Valley. Photograph by Carl Cater

The migration from remote mountain areas to towns in the district and larger cities is a common phenomenon, supported by remittance money; we discuss this issue more in following sections. Currently, one of the greatest challenges to carrying out any entrepreneurial activity is the lack of a skilled workforce (Nyaupane et al., 2019). This is further challenged by the increasing out migration as many young Nepalese go to the Middle East for work (Baruah & Arjal, 2018), which means a shortage of labour capacities for farming and other local livelihoods. The impact of workforce migration is severe in that it changes the demographics and the socioeconomic fabric of mountain communities in Nepal.

Physical capital

As described earlier, physical capital includes infrastructure, which is fundamental to enabling livelihoods and development, such as transport, shelter, water, energy, and communication technologies. The remoteness and terrain of Dhorpatan leads to significant shortcomings in physical infrastructure. Roads into the region are unsurfaced, and the only key access point to the valley is via a steep road from the south that ascends several thousand metres from Burtibang, the nearest town. To the east the road ends at Darbang, several days walk from the valley. As noted by pro-poor tourism advocates, development such as tourism "needs to be complemented by the development of infrastructure… a balanced approach is critical, if transport systems do not exist, the industry will decline and so will any pro-poor strategy" (Gascón, 2017, p. 974).

Within the valley itself, there are vehicle tracks on both the north and south sides, although these are in a poor state. There is no direct link between these two axes, as the river flows along the middle of the valley floor. Vehicles must therefore ford the river at shallow points which vary according to conditions. There are three fixed foot bridges in the valley (which are also accessible by horse/mule, bicycle or small motorbike), however one of these lacks ramp access. A wide number of footpaths link the various settlements in the valley, but these are seemingly neither planned nor well maintained. An airstrip was in use for some years but is derelict and currently not used for fixed wing aircraft operations. Transport infrastructure has long been recognised as a major challenge for poverty alleviation in Nepal, as noted by Blaikie et al. (2002) in prior notes;

> The failure of motorable road provision, as the major form of infrastructural development funded by external investment, to generate a significant or systematic response in terms of agricultural output or diversification, in a context where processes in the wider political economy were not conducive to a capitalist transformation of agriculture.
>
> (p. 256)

Further, most touristic activity will require accommodation infrastructure such as lodges or hotels, which are currently very limited (Nyaupane et al,

2019). Currently there is a campsite near the park office, one guesthouse near the site of the old airstrip that is suitable for the trekking market (shared facilities), and some homestays. However, a mountain bike guide from Pokhara noted that their simple level of culinary provision was not suitable for international tourists requiring complex and high energy food for active pursuits. There is scope, in addition to the usual opportunity for trekking to expand into more diverse adventure-based tourism activities such as mountain biking, paragliding, horse riding, and snow skiing. However, these activities require the development of infrastructure, for example trails and shops.

Energy provision in the region is facilitated by a hydroelectric power scheme which serves the north side of the valley. Excess water from this scheme is fed into a water powered grain mill, a facility built with the support of USAID. There have also been recent developments to explore green energy development, in particular solar energy. Enterprise infrastructure for agriculture is also a key challenge, mainly the lack of modern farming machinery and connections to markets (Nyaupane et al, 2019). This applies to service sectors as well. As mentioned above there are plans for a very large hydropower project. This will involve the construction of a 200m high dam to provide a 821MW plant at the cost of US$1bn (Kathmandu Post, 2019). Changes to hydrology will undeniably have significant positive and negative impacts in the area, but will also submerge a large portion of the valley itself, hence altering livelihoods dramatically.

Certainly, the agenda of Dhorpatan Cluster itself and other stakeholders in the region might push for the improvement of better-connected public infrastructure (Nyaupane et al, 2019). For instance, there are developments externally that may improve access over time, such as the ongoing mid-hill highway construction; although this itself is a barrier to transportation. Internet accessibility is improving and there is mobile coverage for the valley on certain networks (e.g., Nepal Telecom). However, health provision is limited in the valley, with the only significant facility being a small maternity post. Indeed, this lack of infrastructure and services poses significant challenges to the valley, while the fragility of the built infrastructure was demonstrated in November 2018 when a large fire damaged 30 houses in Navi village (Kathmandu Post, 2018). Fortunately, many residents had migrated to lower levels for the winter. Nevertheless, there was significant damage to property and assets (for example stored grains), an estimated cost of Rs 32.8 million was lost.

Financial capital

Financial capital involves the financial resources available to people such as savings, credit, and remittances, the latter being particularly important in Nepal. In pro-poor tourism research, it has been recognised that the greatest poverty alleviation over the long term comes from the dynamic effects of an enlarged economy (Mitchell & Ashley, 2010). However, there is limited formal financial infrastructure within the valley. Although commercial banks are expanding their branch services, this does not seem to be well developed in the Dhorpatan valley itself. Money may be accessed through remittance, micro-loans, and local cooperative and commercial banking. For instance, in 2015, the French charity Amis de Dhorpatan was formed to help farmers grow and sell locally farmed potatoes. Local community owned entrepreneurial ventures as well as investment partnerships such as these are created with an intention to "ensure growing returns and an increasing salary to the producers from one year to the other…" (Amis de Dhorpatan, 2020).

Although entrepreneurial activities have substantially increased in recent years it has failed to create sustainable wealth in the community. It is also typical for organisations, whether for-profit or non-profit, to collapse and become bankrupt within years of their founding. Most businesses in the Dhorpatan region are small grocery retailers connected to the basic trading system in Nepal (Nyaupane et al., 2019). There are limited cottage industries: dairies, livestock, handicrafts, and a few others providing basic services such as IT and banking.

Currently a staggering one third of Nepal's entire GDP comes from remittances (money sent back to family/relatives in Nepal) from overseas workers (Shrestha, 2018). This money is usually spent on land and house purchases which leads to the commercialisation of land (Sunam & McCarthy, 2016). The consequences of remittance money on subsistence farming include lack of stable financial support and a shortage of labour-force, also affecting other precarious livelihoods. Similarly, it appears that the Tibetan community within the valley are also supported by remittances from outside (e.g., a new temple is under construction from foreign funds).

Social capital

Social capital involves resources such as networks, membership of groups, and relationships of trust upon which people draw in pursuit of their livelihoods (Tao & Wall, 2009). Trust in private business as well as the NGO sector has traditionally been very low in Nepal, thus generally it is a challenge to achieve social, environmental, and economic goals despite positive intentions of some actors (Nyaupane et al., 2019). At present, NGOs provide diverse support to the community, including healthcare, education, livelihood, environmental service, youth empowerment and so forth. Though the local government should predominately be responsible for the function of providing such public services, welfare, and other programmes, there has been a history of poor governance as well as a lack of local support to industries and networks (Nyaupane et al., 2019). In response, the Dhorpatan Cluster was created to operate in the Greater Gaumukhi and Dhorpatan region in Baglung, Gulmi and Pyuthan districts in western Nepal. This multi-stakeholder cluster (e.g., community, enterprises, NGOs, government units) is a fairly new asset to the valley, and the region at large. Focused on poverty alleviation, ecology, and enterprise, one of the current Dhorpatan cluster projects underway is the 'Dhorpatan Mountain bike Development Project' which supports running of a shop, events, and skills development. Additionally, as described earlier, local community owned entrepreneurial ventures as well as international partnerships were initiated to support the region (cf. Amis de Dhorpatan, 2020). Further the Federation of Nepalese Chamber of Commerce and Industry (FNCCI) introduced a "one village one product" (OVOP) scheme in the region (and broadly in Nepal), which has been partly successful. The commercial farming of cardamom in the southern region of Dhorpatan is one positive but a limited example.

Cultural capital

Cultural capital, consisting of community heritage, customs and traditions should be considered in any analysis local livelihoods assets (Cater & Cater, 2007). Dhorpatan valley has a mix of cultures and castes including Bishowkarma, Magar, Nauthar (Chhetri), Chhantyal, Thakali, Bhotias (Tibetan refugees), and Brahmin. People are dependent on traditional agricultural practices in the southern part of the hunting reserve. In the northern, higher-elevation parts of the reserve, animal husbandry and trans-boundary trade are the major sources of livelihood

Religious sites are important in the valley, as described above a new Tibetan Bhuddist temple is being constructed in Pathathar Village. At the eastern end of the valley is a Bon temple from an earlier religious sect, while there is a Hindu Uttarganga Baraha shrine on the southwestern slopes of the valley. On the road from Burtibang, just prior to the pass that enters the valley is a religious site located around a small lake. This is popular with local pilgrims and domestic visitors who may visit the site for picnics and to escape the summer heat at lower altitudes.

Political capital

Political capital suggests that there is an empowering role of resources used by actors for the realisation of outcomes that advance actors' perceived interests (Wang, et al., 2016). Political structures are deeply entrenched in Nepal and often form a significant barrier to livelihood development. As Nyaupane and colleagues (2019) highlight,

> The political cadre culture is prevalent everywhere in cities, towns and rural villages… socio-economic life is segregated, dominated and driven by political affiliation and favour as much as certain regions are known to be the "belts" of certain political parties.
>
> (p. 10)

This political structure forms a barrier to development, but at the same time it is important to recognise and work within existing constraints and opportunities frameworks.

Pro-Poor Mountain Tourism Potential in Dhorpatan

Despite some of the challenges highlighted above, Dhorpatan offers significant potential for pro-poor tourism development. Yet there has been little done to promote tourism in the region to date. There have been a small number of promotions in the form of big festivals (Mahotsav) and events organised by charities or communities as well as smaller, regular Hindu festivals which may last several days. Whilst such destination festivals have been able to attract visitors from near and far, the impact is almost immediately lost due to their temporary nature. One of the reasons behind such temporariness is that destinations such as Dhorpatan are very remote and challenging to access as indicated in our assessment of physical capital. Secondly, these rural remote destinations are not well connected to other known national destinations, for example the Annapurna Conservation Area, which is directly adjacent (Nyaupane et al., 2019).

As a mainstay of the Nepalese tourism product, trekking tourism offers opportunities for the communities of Dhorpatan to develop further (Bhattarai, et al, 2005; Nepal, 2003, 2005). Trekking has a long history in the region, principally as an access point to the Dolpo area north of Dhorpatan. Dolpo was mentioned and made popular in the cult book *The Snow Leopard* by Peter Matthiessen (1978). The continued popularity of this adventurous volume amongst trekkers and outdoors people may be a useful opportunity to promote the area, similar to the ways in which books like John Krakauer's (1997) *Into Thin Air* and James Hilton's novel *Lost Horizon* (1933) contribute to the tourism imaginaries of other parts of Nepal (Miller, 2017).

Dhorpatan is also part of the lower Great Himalaya Trail (GHT) which is a long-distance trail crossing the entirety of Nepal, west to east. A recent development has been the less-known Guerilla trek in the west, officially launched in 2012, a heritage trail built in the Maoist conflict heartland. This follows the route and location of significant battles in the civil war of the 2000s. Dhorpatan valley was a training ground for Maoists who stormed the town of Beni in 2004, and is part of the trail which leads from Sulichaur (Rana & Bhandari, 2018). Despite the development of these walking products numbers are still small in comparison to the more popular trekking areas of Annapurna, Langtang, and the Khumbu.

In 2019, our own scoping visit and collaborative explorations with staff from a Pokhara Mountain Bike Company and other stakeholders (eg., Dhorpatan municipality, community members, etc.) identified potential for mountain bike trails in the region. However, there is currently no local bike hire supplier, with quality equipment no closer than Pokhara (at least 8 hours' drive). However, it is likely that Dhorptan would also face competition from much more attractive high-end mountain bike regions. For example, fly-in, fly-out mountain bike tourism is now well developed in lower Mustang, which has the supporting infrastructure that this niche expects.

Although being an internationally known hunting reserve, actual numbers of these niche tourists are very low, approximately 25 per year (DNPWC, 2019). The Department of National Parks and Wildlife Conservation (DNPWC) has allocated 26 blue sheep as an annual quota for hunting in DHR. The quota was never fully utilised in any year between the period of 1980 to 1998, and after this period hunting declined during the civil war. There is still some potential for this market with local involvement as Karki and Thapa (2011) note:

> Involvement of local people in the management of sport hunting is essential. Therefore, to increase their stake, the Reserve and the outfitters should initiate schemes that provide financial support to the concerned local communities. They should also assist local communities to use funds appropriately both in conservation as well as in income generating activities
>
> (2011, p. 30)

However, it is also important to note that the *hunting* title in the protected area designation and brand (i.e. Dhorpatan Hunting Reserve) is a significant deterrent to international tourism. It would be sensible for the reserve to adopt a conservation nomenclature rather than "hunting", as this would allow hunting to continue while simultaneously affording other forms of tourism to be developed and marketed.

Whilst Dhorpatan's adventure tourism appeals to foreign tourists, there is perhaps greater potential for domestic tourism in the region. As discussed above, festivals and other products such as concerts, cultural/religious events and sports competitions could be targeted at local youth and tourists (Nyaupane et al., 2019). These, together with heritage trails in the southern region of the DC, offer rich experiences for local tourism. Many of these may also appeal to international markets, as there is no product that is purely for locals or purely for foreigners, thus they can be blended according to consumer demands.

Finally, there may also be opportunities for agricultural tourism. The Dhorpatan region is rich with good quality high-altitude farming land at between 2000 to 3000 metres, which presents opportunities for specialist crops (such as buckwheat, cardamom and kodo millet), as well as fruits such as oranges and apples. The French charity Amis de Dhorpatan working in the Dhorpatan valley has identified potatoes grown there to have special culinary qualities, with the potential for export (Amis de Dhorpatan, 2020).

Conclusion

It is clear that the Dhorpatan valley faces significant development challenges, whereby local entrepreneurial activities and pro-poor livelihood options lag behind other parts of Nepal. In part this may be due to remoteness of mountain regions themselves and the decade long civil war, but more poignantly is the limited financial, human, and political capital as well as the poor state of physical assets that continue to challenge the communities who reside in Dhorpatan valley. However, a change of government in 2017, and the resurgence of local NGOs and their various projects, signify hope – a sign towards greater economic and social progress in Dhorpatan's future. Moreover, national tourism campaigns (e.g., Visit Nepal 2023 #LifetimeExperiences) pave the way for both domestic and international tourism to be supported by, as well as potentially augment the high levels of environmental, social, and cultural capital that already exists in the Dhorpatan area.

While this shiny promise of tourism begins to be touted within the valley, these types of development will require local collaboration, dialogue and some sort of consensus among the various stakeholders in the region (Nyaupane et al., 2019; see also Apollo & Wengel, 2022), particularly if it is to benefit the poorest. Indeed, a note of caution has been recognised by Mitchell and Ashley (2010) who contend that "just because a tourism segment is built on wildlife or culture does not mean it is pro-poor" (p. 134). There is clearly a need to further research the issues faced by residents in the region and provide using diverse methodologies that can more accurately portray their individual views on their assets. Ongoing research intends to conduct more detailed work with community stakeholders, such as mapping exercises (Cater, 2012) to identify community aspirations. Furthermore, these stakeholders and communities are urged to conceptualise tourism as part of the solution, not the saviour. Here, tourism becomes one of the sustainable and alternative livelihoods that may afford local economic activities that can offer opportunities (skilled and well-paid jobs) that enable migrating youths to stay in Nepal, and disrupt increasing out migration trends. This proposition of course needs additional research. Yet our pro-poor sustainable livelihood approach has, perhaps, played a small part in helping us begin to address such critical sustainability challenges.

Management Implications

Our examination of Dhorpatan shares much with pro-poor mountain challenges in other parts of the world. Whether hosting tourists or not, peripherality is the challenge that all mountain regions face. Whilst the promise of improved infrastructure such as roads or dams may bring development, it is a difficult balancing act to maintain a clean and beautiful environment whilst also promoting the economic and other interests of local mountain dwellers and the various vested stakeholders, as suggested by this book. Thus, it is highly encouraged that tourism developers and managers engage in local dialogue and collaboration with various stakeholders in the region to identify not only local interests, but assets as well. It is through a sustainability approach such SLA that solutions may be realized for some of the demographic and socio-economic challenges experienced in mountainous regions. For instance, where agriculture is often the main means of livelihoods and commerce, communities could consider combining agriculture and nature-based tourism as an opportunity. While tourism infrastructure may need enhancement (as we saw in Dhorpatan) these community facets can still play a small role in supporting local employment and reducing outmigration.

It should be noted that highly centralised governance administration system (such as that in Nepal up until the late 2010s) can often exacerbate the problems we see in remote mountainous regions. Thus, it's suggested to seek joint initiatives through non-profits and social enterprises (with local engagement) and multilevel and multisectoral funding to help support sustainable tourism and local economic development. Relatedly, the development and management of connected trails and tourism clusters are important initiatives, which can help to position remote destinations in a crowded tourism marketplace. However, as noted earlier, the most significant pro-poor tourism impacts come not with the direct or indirect economic impacts of tourism (which in mountain regions is likely to be small scale anyway), but indeed with the induced effects of a larger, more diversified economy making the most of its resource base. Thus, we encourage managers and developers of remote mountain destinations to identify these resources as their first step towards more sustainable, nature-based livelihoods in mountain regions.

References

Amis de Dhorpatan. (2020). *Friends of Dhorpatan – Potatoes from Nepal.* Retrieved from: http://amisdedhorpatan.org/en/welcome/#the-association.

Apollo, M. (2017). The true accessibility of mountaineering: The case of the High Himalaya. *Journal of Outdoor Recreation and Tourism*, 17, 29–43.

Apollo, M. & Wengel, Y. (2022). *Mountaineering Tourism: A Critical Perspective.* London: Routledge.

Ashley, C. (2000). *The impacts of tourism on rural livelihoods: Namibia's experience*, working paper 128. Overseas Development Institute: London.

Ashley, C. & Roe, D. (2002) Making tourism work for the poor: Strategies and challenges in southern Africa. *Development Southern Africa* 19(1), 61–82.

Baruah, N. & Arjal, N. (2018). *Nepalese Labor Migration – A Status Report.* The Asia Foundation. [Online]. Retrieved from: https://asiafoundation.org/2018/06/06/nepalese-labor-migration-a-status-report/.

Bhattarai, K., Conway, D., & Shrestha, N. (2005). Tourism, terrorism, and turmoil in Nepal. *Annals of Tourism Research*, 32(3), 669–688.

Blaikie, P., Cameron, J., & Seddon, D. (2002). Understanding 20 years of change in West-Central Nepal: Continuity and change in lives and ideas. *World Development*, 30(7), 1255–1270.

Cater, C. (2012) Community involvement in trekking tourism: The Rinjani Trek Ecotourism Programme, Lombok, Indonesia. In Garrod, B. and Fyall, A., *Contemporary Cases in Tourism* (pp191–212, Vol 1). Goodfellow.

Cater, C. and Cater, E. (2007). *Marine ecotourism: Between the devil and the deep blue sea.* CABI: Oxfordshire.

Cater, C., Garrod, B. & Low, T. (eds.) (2015). *The Encyclopedia of Sustainable Tourism.* CABI: Oxford

Department of National Parks and Wildlife Conservation (DNPWC) (2019) Home page Retrieved from: http://www.dnpwc.gov.np/en.

Gascón, J. (2017). Pro-poor tourism. In Lowry (ed), *The SAGE International Encyclopedia of Travel and Tourism.* Vol. 4. Thousand Oaks, CA: SAGE Publications, Inc.

Helmore, K. & Singh, N. (2001). *Sustainable livelihoods – Building on the wealth of the poor*. Kumarian Press: Bloomfield.

Hilton, J. (1933). *Lost Horizon*. New York, NY: William Morrow & Company, Inc.

Kandel, R. (2000). Status Paper of Dhorpatan Hunting Reserve. *Grassland Ecology and Management in Protected Areas of Nepal*, 3, 137–145.

Kathmandu Post. (2018). Fire destroys 30 homes in Baglung. November 26, 2018 https://kathmandupost.com/national/2018/11/26/fire-destroys-30-homes-in-baglung.

Kathmandu Post. (2018). Six international firms vying for 821MW Uttar Ganga consulting contract. November 5th2019. https://kathmandupost.com/money/2019/11/05/six-international-firms-vying-for-uttar-ganga-consulting-contract.

Karki, J. B., & Thapa, B. B. (2011). Status of blue sheep and Himalayan tahr in Dhorpatan Hunting Reserve, Nepal. *Banko Janakari*, 21(1), 25–30.

Krakauer, J. (1997). *Into thin air*. New York, NY: Villard Books.

Lee, M. H. (2008). Tourism and sustainable livelihoods: the case of Taiwan. *Third World Quarterly* 29(5), 961–978.

Mair, H. (2006). Global restructuring and local responses: Investigating rural tourism policies in two Canadian communities. *Current Issues in Tourism*, 9(1), 1–43.

Matthiessen, P. (1978). *The Snow Leopard*. Vintage.

Miller, M.C. (2017). *An exploration of Sherpa's narratives of living and dying in mountaineering*. (Doctoral dissertation, University of Waterloo, Canada). Retrieved from: https://uwspace.uwaterloo.ca/handle/10012/11895.

Ministry of Culture, Tourism & Civil Aviation. (2020). *Nepal Tourism Statistics 2019*, Government of Nepal Ministry of Culture, Tourism & Civil Aviation. Singha Durbar, Kathmandu. Retrieved from: https://www.tourism.gov.np/files/.

Mitchell, J., & Ashley, C. (2010). *Tourism and poverty reduction: Pathways to prosperity*. Taylor & Francis.

Nepal, S.K. (2003). *Tourism and the environment: Perspectives from the Nepal Himalaya*. Nepal: Himal Books.

Nepal, S.K. (2005). Tourism and remote mountain settlements: Spatial and temporal development of tourist infrastructure in the Mt. Everest Region, Nepal. *Tourism Geographies*, 7(2), 205–227.

Nyaupane, R., Pretious, M., and Khadka, S. (2019) *Issues in the Development of Community-led Tourism in Nepal: the example of Dhorpatan region*. Presented at the Annual Kathmandu Conference On Nepal And The Himalaya, July 2019, Kathmandu.

Panthi, S., Aryal, A., Raubenheimer, D., Lord, J., & Adhikari, B. (2012). Summer Diet and Distribution of the Red Panda(Ailurus fulgens fulgens) in Dhorpatan Hunting Reserve, Nepal. *Zoological Studies*, 51(5), 701–709.

Rana, S., & Bhandari, C. (2018). Guerrilla Trail: A tourism product in the Maoist rebel's footsteps. *Journal of Tourism*, 19(1), 33.

Scheyvens, R. (2007). Exploring the tourism-poverty nexus. *Current Issues in Tourism*, 10(2–3),231–254.

Scoones, I. (1998). *Sustainable rural livelihoods: A framework for analysis*. University of Sussex, IDS: Brighton.

Shrestha, S. (2018). *The plight of Nepal's migrant workers*. Al Jazeera. [Online]. Retrieved from: https://www.aljazeera.com/blogs/asia/2018/02/plight-nepal-migrantworkers-180205084756399.html.

Sunam, R. K., & McCarthy, J. F. (2016). Reconsidering the links between poverty, international labour migration, and agrarian change: Critical insights from Nepal. *The Journal of Peasant Studies*, 43(1), 39–63.

Tao, T.C.H. & Wall G. (2009). A livelihood approach to sustainability. *Asia pacific Journal of Tourism Research* 14(2), 137–152.

Telfer, D.J. & Sharpley, R. (2016). *Tourism and development in the developing world.* Routledge.

Thapa, B. B., Panthi, S., Rai, R. K., Shrestha, U. B., Aryal, A., Shrestha, S., & Shrestha, B. (2014). An assessment of Yarsagumba (Ophiocordyceps sinensis) collection in Dhorpatan hunting reserve, Nepal. *Journal of Mountain Science*, 11(2), 555–562.

United Nations Development Programme. (2023). Human development reports. *United Nations Development Programme.* Retrieved from http://hdr.undp.org/en/countries/profiles/NPL.

Wang, J, Cater, C. & Low, T. (2016) Political challenges in community-based ecotourism. *Journal of Sustainable Tourism*, 24(11), 1555–1568.

World Bank (2023). *Data: Nepal population.* The World Bank. Retrieved from http://data.worldbank.org/country/nepal.

3 The Genesis of Pro-Poor Mountain Tourism in Rural China

Three Cases in Enshi Prefecture

Huafang Qiao, Rick Gage, Lina Xiong and Shuangyu Xie

Introduction

Mountain landscapes have drawn adventurers for centuries, but only in recent decades has recreational tourism gained a foothold in remote highland communities (Apollo & Wengel, 2022; Macfarlane, 2004; Messerli & Ives, 1997). While tourism has acted as a vehicle for economic growth in mountainous regions around the globe, these unique environments and cultures appear increasingly under threat amid growing global tourism pressures (Nepal, 2000; Nepal & Chipeniuk, 2005). Issues around soil erosion, water pollution, vegetation loss, and wildlife impacts have been reported even from relatively low-impact activities like hiking (Apollo, 2017; Huddart & Stott, 2019). In addition, rapid tourism expansion exacerbates risks of environmental degradation, cultural disruption, and unequal local impacts if not paired with coordinated planning and protections (Apollo, 2017; Nepal, 2008). Managing this strain between economic livelihoods and ecological integrity represents an urgent challenge, especially across fragile, mountainous regions (Guo, et al., 2022; Neba, 2010). Alternatively, mindful development enables rural communities to capitalise on spectacular mountain assets as a pathway out of poverty (Marek & Wieczorek, 2015; Neba, 2010; Zurick, 1992). This chapter examines the emergence of mountain tourism in Enshi Prefecture, China, through case studies of recent development projects oriented toward poverty alleviation.

China is one of the most mountainous countries in the world, and its mountain regions are home to some of its poorest citizens. In 2011, the Chinese government established goals to eliminate abject poverty nationwide by 2020. In rural mountainous areas, tourism has been a primary driver of these efforts. China's policy support for pro-poor tourism initiatives has facilitated inclusive economic gains for previously isolated areas, though outcomes vary (Qiao & Xie, 2022; Xie et al., 2020). Evaluating these successes and pitfalls provides practical insights to inform policy and practice in alignment with principles of sustainable development (Beedie, 2003; Beedie & Hudson, 2003; Jia et al., 2021).

This analysis focuses on emerging cases of small-scale tourism growth tied to poverty alleviation efforts in mountain villages of Enshi Prefecture, an ethnically diverse region of western Hubei Province, China. Until the early

DOI: 10.4324/9781003392583-4

2000s, this remote and isolated region was largely excluded from modern development. As such, it provides a unique look into the beginning phases of tourism development and the associated impacts on the natural, socio-cultural, and economic environments. Advancing pro-poor approaches in culturally and ecologically fragile settings remains contingent on deliberate governance measures alongside market-based incentives (Cai & Wu, 2018; Nepal, 2008; Zurick, 1992). By examining projects spearheaded by different combinations of government entities, private developers, and local councils, this chapter reveals distinct models with strengths and limitations in the fragile mountain environment. Understanding the complex dynamics that promote sustainable livelihoods and/or environmental exploitation can guide properly balanced decision-making as mountain tourism expands globally (Chambers & Conway, 1992; Wang et al., 2019).

Background: Pro-Poor tourism in Enshi Prefecture, China

Pro-poor tourism in China has evolved through four distinct stages since the late 1970s (Qiao & Xie, 2022). Initially, during spontaneous development (1979–1990), tourism enterprises emerged without much planning, proving beneficial for local economies. The government recognised tourism's potential in the late 1980s, including it in the national plan. The development exploration stage (1991–1999) saw formal recognition of tourism for poverty alleviation, gaining support at national and local levels. The pilot development stage (2000–2010) witnessed substantial investments in large-scale projects, solidifying tourism as a social and economic development tool. Since 2011, the government has intensified efforts in tourism poverty alleviation, culminating in the successful achievement of national poverty alleviation goals by the end of the decade, with tourism playing a pivotal role (Guo et al., 2022; Qiao & Xie, 2022; Zhao, 2017).

Enshi Tujia and Miao Autonomous Prefecture (Enshi Prefecture), situated in the mountainous southwest of Hubei Province, stood as one of China's most impoverished regions before tourism development. Comprising eight municipalities, including Enshi City, the area boasts stunning natural landscapes and a rich cultural heritage, particularly of the Tujia and Miao ethnic minorities, making it an attractive destination for mountain and cultural tourism (Han et al., 2019). Despite its scenic beauty, the region faced persistent poverty due to its remote location and limited infrastructure. Following the central government's guidance, Enshi Prefecture responded by embracing tourism as a crucial strategy for poverty reduction, employing a participatory approach to involve local communities in decision-making and planning (Guo et al., 2022; Tao & Wall, 2009). Residents actively participate in tourism development through community-based organisations, contributing their knowledge, skills, and resources. Training programs focused on hospitality, guiding, handicraft production, and sustainable resource management have been implemented to empower the local population. Inclusive business models, including homestays and handicraft cooperatives, have been encouraged, fostering entrepreneurship and reducing dependence on external assistance. Despite challenges, tourism emerged as a leading industry, contributing 46% of the annual GDP in 2019, showcasing the success of Enshi Prefecture's approach to sustainable and community-driven tourism development (Statistics Bureau, 2020). The diverse development models and distinct tourism resources require further examination at the local destination level.

Methodology

This chapter employs an exploratory multiple case study approach to examine emerging pro-poor mountain tourism development models in Enshi Prefecture. Data were collected through semi-structured interviews with key informants during site visits to three tourism destinations representing different development approaches: Enshi Grand Canyon, Tenglong Cave, and Pingshan Scenic Area. The locations of the three case sites are shown in Figure 3.1. Additional data sources include government reports, media accounts, and scholarly literature.

At each site, interviews were conducted with tourism company executives, local government officials, village leaders, and residents. Interview questions focused on understanding the development history, business models, resident participation, poverty alleviation initiatives, and environmental protection efforts associated with each of the three destinations. In total, approximately

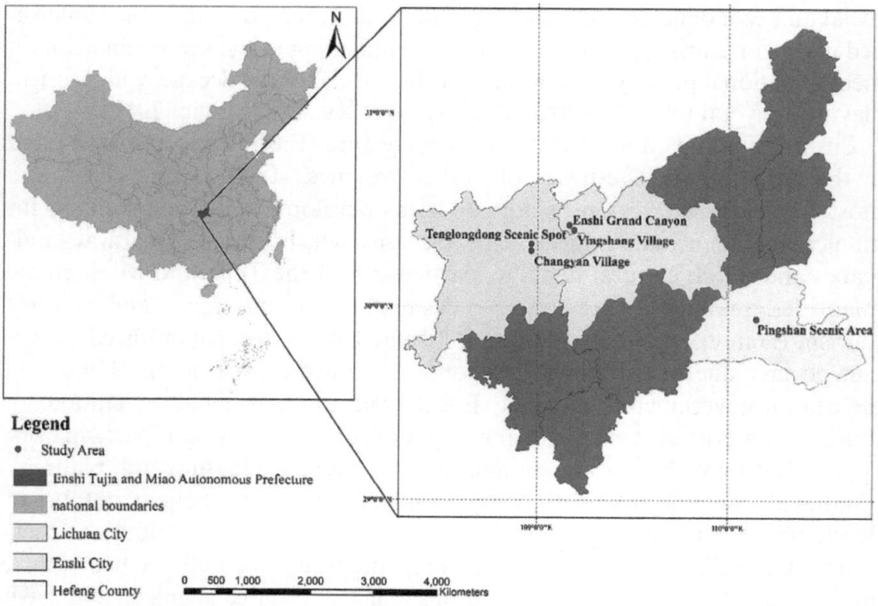

Figure 3.1 Map of the study area

30 interviews were conducted across the three sites. Interview data were analysed using an inductive coding to identify key themes related to tourism impacts, sustainability, and local livelihoods. In addition, detailed observations were made during site visits to document infrastructure development, commercialisation within attractions, waste management practices, transportation access, participation of local community members, and other relevant aspects. Secondary data from government statistics, company reports, media coverage, and existing literature were also consulted to provide background information and supplement findings from interviews and observations.

This exploratory study provides an in-depth look at the development of Enshi Grand Canyon, followed by summarised overviews of Tenglong Cave and Pingshan Village. The case of Enshi Grand Canyon showcases a state-owned tourism company's large-scale development of a prominent natural attraction, including extensive infrastructure construction, resident participation models, poverty alleviation initiatives, and environmental protection efforts. The subsequent cases highlight alternative development approaches led by an external state-owned tourism company in partnership with the local government (Tenglong Cave) and a local private tourism enterprise working alongside the county government (Pingshan Village). Similarities and differences across the three cases are noted throughout the results, focusing on tourism investment models, resident participation, poverty alleviation measures, and environmental protection. While Enshi Grand Canyon is explored extensively as an instrumental case study, including comparative context-specific findings from the other two sites, it provides a more comprehensive perspective on the region's emerging models of pro-poor mountain tourism development.

Results

In-depth Case Study: Enshi Grand Canyon

Overview

Enshi Grand Canyon is one of the world's longest and most dramatic canyon landscapes. The canyon is over 108 kilometres long and covers over 300 square kilometres (Figure 3.2). It boasts a staggering display of complicated karst geomorphology, including caves, sinkholes, stone pillars, natural bridges, fissures, and sheer cliffs. Over 90% of the area is covered by forests, and the climate is generally mild, with no harsh winters or summers. The Yunlong Gap that cuts through the canyon has been dubbed "the most beautiful scar of the earth", and the area has appeared on numerous lists of the most scenic places in China. Before tourism development, most roads were dirt and gravel and filled with potholes and other hazards. Living conditions, such as medical and education facilities, were poor due to the lack of roads. Villagers generally needed to walk a whole day to get supplies (e.g., fertilisers) from outside the mountains.

Figure 3.2 Attractions in Enshi Grand Canyon (outside view, hanging pathway, unique rock formation)

In July 2015, Enshi Grand Canyon was designated as a National 5A Tourist Attraction (the highest level) by the National Tourism Administration, and in 2018, it was listed as one of the top ten most influential attractions in Hubei Province and No. 3 on the top 50 most influential national 5A tourist attractions in China. In 2019, the area hosted more than 1.5 million tourists and was named a must-see attraction in Enshi. In 2018, the World Tourism Alliance recognised Enshi Grand Canyon as a remarkable case of poverty alleviation through tourism.

One of the most remarkable things about this region is that it was virtually unknown to the outside world before the turn of the century. In 2003, an early expedition team introduced the unique karst landscape to the world, and in 2004, a joint Sino-French karst cave research team began exploring the region in depth. It has only been open to the public since October 2010. In 2005, the Enshi

Municipal Government and a foreign-funded private enterprise (the Huangshan Management Committee) signed an investment and development agreement. In 2010, Hubei Travel and Investment Company (HBTI), a state-owned development company, purchased the contract and began investing in tourism infrastructure. In 2013, HBTI changed its name to Hubei Culture & Tourism Group Co., Ltd.(HBCTG) and began incorporating local Tujia and Miao cultural practices, architecture, art, and folklore into their development projects.

Development model

Enshi Tourism Group is a subsidiary of HBTI (now Hubei Culture &Tourism Group Co., Ltd.– HBCTG). The provincial government established this subsidiary to push for strategic tourism development that focused on local natural and cultural resources. Since its establishment in October 2010, they have been actively directing the development and management of Enshi Grand Canyon. A goal was set in 2012 to become a 5A tourism attraction (a goal that was accomplished in only three years), and the company immediately began constructing visitor centres, tourist cable cars and lines, vertical sightseeing elevators, elaborate hanging platforms and pathways along the sheer rock face, and much more. By 2019, the company had invested over two billion RMB into development projects and committed over one billion more over the next few years. Along with the rapid tourism development, there has been tremendous growth in transportation infrastructure fuelled by governmental efforts, including the extension of a high-speed railway line into Enshi and drastically improved roadways to accommodate the influx of visitors.

Resident participation in tourism

With tourists coming in droves, local villagers have increasingly participated in tourism and profited from the development boom. Today, tourism development has become the main economic driver for the local village, with 80% of the park workers originating from the surrounding villages. In total, this state-owned company employs approximately 800 workers in over 200 shop booths and service positions, as "sliding pole" carriers (a type of chair attached to a stretcher on which visitors can be carried through the canyon), restaurant and lodging employees, maintenance workers, or in other roles. Some of the most lucrative positions are operating the six shop booths at the top of the mountain in the "Golden Zone." An annual lottery system distributes shop booths, and those drawn to operate a booth pay the company a yearly fee. For all shops, the fee is a meagre 2000 RMB per year (approximately USD 275). However, for booths in the Golden Zone, the shop operators must also pay a fee of 120,000 RMB (approximately USD 16,500) each to the village committee. The village committee will then distribute the pooled money to other families equally. HBCTG sets prices for goods sold at the shops and is not to be changed by shop owners (although there is some flexibility with lowering prices). This guarantees that the highest prices can be charged. The annual income for most shop workers is 50,000–60,000 RMB but is considerably higher for those in the Golden Zone.

Residents' participation in tourism also extends beyond the scenic area's boundaries. More than 300 farm-to-table and bed and breakfast style lodging services have over 3000 beds and 4000 table seating capacity. Some well-operated lodging businesses can bring in half a million RMB (about USD 70,000) in annual household income. Other income sources are agriculture (corn, potatoes, tea, and medical herbs), foraging for mushrooms and wild vegetables, and migrant labour. Those closer to the canyon tend to reap the financial benefits of tourism more than those farther away, but the benefits from improved infrastructure extend to nearly everyone. In general, villagers are highly supportive of new construction and development projects due to tourism.

Land acquisitions

In the development of Enshi Grand Canyon, HBCTG signed a land use contract with the local village committee as the collective landowner. The contract dictates that HBCTG can use the land because the village receives a five per cent share of the business. HBCTG also signed a contract with the village committee guaranteeing priority employment of local villagers, especially the ones who have contracted out land to the company. Around 3000 mu (approximately 500 acres) of land was acquired by the company, and the money was paid to the village collective and distributed to villagers affected by the agreement.

Poverty alleviation

The local village councils have adopted several poverty alleviation measures related to tourism development. In 2016, they established a farm business association that coordinates a local supply chain, offers training and guidance to farmers, and encourages employing individuals in poverty. They also initiated a program to evaluate local tourism businesses (farm-to-table restaurants and lodging) and issue cash awards to top performers. They offer tourism service training to business owners and participants (chefs, lodging operations, etc.) and organise field trips outside the village for additional training. They also invested 3.6 million RMB in updating houses within the attraction, including painting exterior walls and updating the roofs for a consistent look. In particular, given that HBCTG is a state-owned company, resident well-being is one of their top priorities. The leadership team of HBCTG is required to join the national effort and work directly with villagers to alleviate poverty. During the COVID-19 pandemic, the attraction was kept open as much as possible, and a heavy economic loss was experienced. However, they did not lay off or cut pay for any employees. They actively participated in the local pandemic prevention and mitigation measures, building a strong connection with local residents.

Environmental protection

To preserve the rich natural resources in the Enshi Grand Canyon area, tourism development prioritises environmental protection and geological disaster prevention. As a state-owned company, HBCTG is expected to prioritise local residents' well-being and protect the natural environment for sustainable development. HBCTG has invested over 200 million RMB in projects to reinforce natural stone structures, secure dangerous areas, build wastewater treatment plants, and institute trash removal programs and facilities. When asked about environmental conditions after the arrival of tourism, several villagers mentioned that there is no garbage on the streets, there are many receptacles around the area, and merchants are responsible for hauling out their own garbage each day to be properly disposed of (as a part of the booth operation agreement).

Comparison of two other cases in Enshi Prefecture

Tenglong Cave (Figure 3.3) is a national 5A attraction developed through a partnership between an external state-owned tourism company and the

Figure 3.3 Attractions in Tenglong Cave (top row) and Pingshan Village (bottom row)

local government of Lichuan City. The site features intricate karst cave systems along with a modern showcasing of Tujia cultural heritage through elaborate song and dance shows. Development has prioritised rapid growth and commercialisation with amenities like restaurants, shops, and a train tour inside the cave complex. Due to the mass tourism development model, tourists rarely spend more than a few hours on-site, limiting economic benefits to the community. Poverty alleviation strategies follow limited approaches of one-time land compensation and student scholarships. While bringing economic benefits, tensions persist around unequal distribution of profits, inauthentic representation of minority culture, and environmental impacts from extensive construction inside the natural cave system.

In contrast, tourism development in Pingshan Village (Figure 3.3) is led by a local private company, Baotong Tourism Development, working closely with the county government. Tourism resources leverage dramatic slot canyon landscapes and traditional Tujia minority cultural heritage. The approach emphasises environmental sustainability by designating 23 square kilometres as a protected no-access preservation zone. Poverty alleviation initiatives focus on annual financial support and incentives for local participation in tourism enterprises like farm-to-table restaurants. However, strict access limits on protected areas have caused tensions, with some villagers desiring more infrastructure development and opportunities to profit directly from tourism expansion.

Discussion

While the in-depth case study of Enshi Grand Canyon illustrates a provincial state-owned company's approach to tourism development, poverty alleviation, and sustainability, comparing and contrasting core elements of this case to the alternative models seen at Tenglong Cave and Pingshan Village provides critical analytical insights. If we consider a spectrum of approaches with development at one end and protection at the other, Enshi Grand Canyon can be thought of as the middle-of-the-road approach, with Tenglong Cave representing a case of heavy commercial development and Pingshan Village representing a case of intensive resource protection. By examining various strategies in tourism, including resource leverage, investment models, poverty relief, and environmental conservation, we can go beyond site-specific findings to propose explanations for the implications of development model choices on sustainability outcomes. The discussion will systematically highlight similarities and differences across cases, identifying patterned variances based on development strategies. This allows us to argue the complex tensions between economic growth, poverty alleviation, heritage preservation, and environmental protection shaping emerging pro-poor mountain tourism models.

Tourism resources

The cases showcase diversity across sites regarding core tourism resources. Enshi Grand Canyon leverages a stunning natural landscape that includes forests, waterfalls, and intricate karst geomorphology. Tenglong Cave also capitalises on a geological wonder as the world's largest cave entrance complex but adds a modern interpretation of local cultural elements with elaborate Tujia dance performances, clothing, and architecture. In contrast, Pingshan Village features dramatic physical landscapes like slot canyons and a more traditional showcasing of the rich cultural heritage as a historically isolated Tujia minority settlement. While the sites all combine natural and cultural assets in some way to attract visitors, they differ regarding primary draw factors and alignment between development and local characteristics.

Tourism investment models

There is also variation in terms of leading tourism investors at each destination. As a renowned natural landmark in Hubei Province, Enshi Grand Canyon has benefitted from major infrastructure investments by a provincial state-owned company focused on scenic area development, local transportation expansion, and highlighting minority cultures. With prized assets but more remote locations, Tenglong Cave and Pingshan Village have adopted blended partnership models combining external state-owned companies or local private capital with local government support.

The level of investment and development orientation differs considerably across cases. However, in general, when ownership and management are guided by external entities (as in the case of Tenglong Cave), development and commercialisation appear to be prioritised more highly than the preservation of local ecological or cultural resources. Conversely, if local development companies and government agencies primarily make decisions with direct ties to the community, more emphasis is placed on preserving the integrity of local characteristics. However, this can lead to tension between village residents and site developers as it can limit the ability of community members to profit from tourism, as has been observed in Pingshan Village.

Poverty alleviation strategies

All three sites have embraced poverty alleviation objectives alongside tourism development, but the specific strategies differ across cases. As a state-owned company, the developer at Enshi Grand Canyon has been most active in incorporating poverty relief programs as core business practices through preferential hiring, profit sharing, skills training, and awards programs. As seen in Tenglong Cave, external investor-driven models follow more limited "equivalence exchange" approaches focused on specific incentives that provide direct returns like one-time land transfer compensation payments and student scholarships. While success has been achieved across cases in boosting resident incomes, there is variance regarding the centrality of poverty alleviation in the development agenda.

Environmental protection measures

Each tourism development entity espouses commitments to environmental sustainability, particularly protecting core natural assets, but tensions persist around growth priorities. Enshi Grand Canyon's state-owned developer has invested heavily in conservation infrastructure, allowing controlled commercialisation. Locally-focused Pingshan Village protects nearly a quarter of the total land area despite economic pressures to expand tourism development. Tenglong Cave's external private investors have taken a lighter approach, implementing water treatment and cooking bans inside the cave but allowing extensive land alterations and construction to boost mass tourism. Approaches to environmental protection in Enshi Prefecture have showcased a continuum from mass commercialisation to restrictive development, with different trade-offs and outcomes experienced.

Summary of cases

As evidenced across the cases of Enshi Grand Canyon, Tenglong Cave, and Pingshan Village, balancing ecological integrity, cultural heritage preservation, and economic prosperity is an immense challenge in developing mountain tourism destinations (Table 3.1). Despite stated sustainability commitments, the varying levels of environmental protection efforts across the sites illustrate persistent difficulties of aligning conservation incentives with growth imperatives. Enshi Grand Canyon's extensive investments in securing hazardous areas and building conservation infrastructure stand out as a model for proactively enhancing resilience, though it is costly and may not be possible for resource-limited communities. The short-term development orientation at Tenglong Cave needs to incorporate more opportunities for community involvement and efforts to facilitate more extended visits and overnight stays. In Pinghan, while their determination to protect their core resource is commendable, more public communication and the development of additional revenue-generating methods are encouraged. Finally, all sites would benefit from more robust visitor education programs focused on responsible behaviour and minimising environmental impact as participation expands (Macfarlane, 2004; Nepal, 2000). Table 3.1 provides a summary of these discussions.

Conclusions

Promising strides in poverty alleviation have been made in the presented cases in Enshi Prefecture, yet sustaining the management of distinctive natural

Table 3.1 Comparisons of tourism cases for key dimensions

Key Dimension	*Enshi Grand Canyon*	*Tenglong Cave*	*Pingshan Village*
Core Tourism Resources	Dramatic natural landscapes including karst geomorphology features like waterfalls, caves, stone pillars.	Massive natural cave complex featuring largest cave entrance in world. Modern cultural performances showcasing Tujia heritage.	Stunning slot canyons, rivers, forests. Traditional Tujia cultural village with rich tangible and intangible heritage.
Development Model	Provincial state-owned company investment focused on infrastructure, transportation expansion, commercial amenities, and cultural preservation within site.	Joint private-public model with external company partnership driving mass tourism approach – significant on-site construction and entertainment amenities development.	Local private company working with county government to develop limited amenities and necessary infrastructure improvements, balancing tourism access and strict resource protections.
Resident Participation in Tourism	Widespread direct employment at site (~800 jobs). Significant involvement in lodging, restaurants, transportation, and other tourism-related enterprises (~300 local businesses). Lottery system for lucrative shop booth operations.	Limited opportunities – some retail permits, labor jobs (~120 employed), expectations of expanded participation through planned infrastructure upgrading surrounding villages. The degree of resident involvement is largely dictated by the company.	Direct employment for some residents (~30 jobs). Encouraged involvement in lodging, restaurants, transportation, sales – supported through training initiatives targeting poor residents.
Land Ownership and Use	Contract with village committee to lease collectively managed lands – compensation provided. Priority hiring for affected villagers.	Company acquires land for project development with financial incentives provided to relocating families.	Land ownership is primarily retained by community residents, while developments are largely funded by the local company and village council.
Poverty Alleviation	Active corporate social responsibility programs – awards for local businesses, skills training and other support infrastructure. 5% profit sharing with collective.	Limited initiatives – incentives mainly tied directly to company's land acquisition and development plans. Expect future indirect benefits from village upgrading initiatives.	Direct financial support for students and disadvantaged groups. Covers operational costs for village committee. Support tied to tourism-related skill building for poor residents.

Key Dimension	*Enshi Grand Canyon*	*Tenglong Cave*	*Pingshan Village*
Cultural Heritage Preservation	Blending of architectural styles, emphasis on Tujia/Miao folklore and ethnic creative culture into site design and performance programming.	Modern take showcasing Tujia song and dance, food and drink, and commercialised souvenirs; less alignment with authentic historical cultural practices and artifacts.	Strict protections on access to historic village zone preserving integrity of cultural assets. Limited display of traditional cultural practices.
Environmental Protection	Significant investments in conservation infrastructure while allowing controlled commercialisation. Visitor communication (directional and interpretive signage) is limited.	Light sustainability focus – water treatment systems installed but highly commercialised alteration of natural cave environment.	Strict protected zone prohibiting access or development impacts nearly a quarter of total land area.
Key Tensions	Managing growth while preserving integrity of natural and cultural assets. Ongoing negotiations around benefit sharing. Difficulty in balancing social responsibility with economic gains.	Inequitable distribution of profits and decision-making power. Loss of authentic cultural heritage. Environmental exploitation concerns. Local resentment related to tourism benefit distribution.	Conflicting pressures from tourism expansion desires from residents and strict sustainability protections by the company. Disagreements over appropriate resource use.

landscapes crucial for regional tourism assets necessitates collaborative governance efforts across sectors (Mayor, 2021; Nyaupane, 2015; Wang et al., 2021). Balancing environmental protections with infrastructure development poses a significant challenge in pro-poor mountain tourism. Proactive planning and ongoing benefits-sharing negotiations can help mitigate risks, particularly when local communities are empowered in decision-making processes (Lo et al., 2021). In Enshi Prefecture, the sustainability of the regional tourism industry hinges on resilient management strategies that conserve vulnerable ecological and cultural assets that attract visitors. These cases illustrate how well-managed pro-poor mountain tourism can enhance socioeconomic outcomes while safeguarding ecological and cultural resources. However, the region's rapid growth underscores the imperative for coordinated policies among government agencies, developers, councils, and residents to ensure long-term sustainability (Nepal, 2008). As exemplified in the Enshi Grand Canyon case, state-owned enterprises play a crucial role in balancing environmental protections with profit-oriented development. Moreover, fostering awareness of sustainable behaviours among tourists and promoters is vital for the longevity of mountain tourism destinations (Beedie, 2003; Nepal, 2000). The unique history and ongoing evolution of rural pro-poor tourism position it as a leader in sustainable tourism development in fragile yet captivating mountain settings.

Management Implications

While the experiences and lessons learned from the presented cases in Enshi Prefecture may be unique to the area, the overall implications include long-term planning that integrates local natural and cultural characteristics and facilitates communities' participation in tourism development. Financial incentives and social support that cater to locals' needs and wants are critical. Ongoing communication, coordination, and positive ties among local communities, governments, and development companies safeguard the sustained growth brought about by tourism. Several practical recommendations can be drawn from this research to guide management approaches in other regions pursuing pro-poor mountain tourism:

- Engage local communities extensively in participatory planning to leverage traditional knowledge and foster empowerment in tourism enterprises. This helps align development initiatives with cultural and environmental resource stewardship.
- Promote community-based tourism models like resident-owned small businesses to maximise local retention of tourism profits while providing training, incentives, and oversight to ensure quality and sustainability.
- Implement robust visitor education programs on low-impact practices and responsible behaviours to minimise tourist pressures.
- Design protected zones prohibiting or limiting development and human activities to conserve the most ecologically and culturally significant locations, recognising that some preservation is vital amid growth.
- Encourage tourism companies and local leaders to invest portions of profits into conservation initiatives to offset environmental impacts and demonstrate long-term sustainability commitments.

While strategies must be tailored to each context, these principles can guide policy-makers and planners toward mountain tourism development that alleviates poverty while respecting ecological limits and cultural heritage.

References

Apollo, M. (2017). The population of Himalayan regions – By the numbers: Past, present and future. In R. Efe & M. Öztürk (Eds.), *Contemporary studies in environment and tourism* (pp. 145–159). Cambridge Scholars Publishing.

Apollo, M. & Wengel, Y. (2022). *Mountaineering tourism: A critical perspective.* Routledge. https://doi.org/10.4324/9781003095323.

Beedie, P. (2003). Mountain guiding and adventure tourism: Reflections on the choreography of the experience. *Leisure Studies*, 22(2), 147–167. https://doi.org/10.1080/026143603200068991.

Beedie, P., & Hudson, S. (2003). Emergence of mountain-based adventure tourism. *Annals of Tourism Research*, 30(3), 625–643. https://doi.org/10.1016/S0160-7383(03)00043–00044.

Cai, J.J. & Wu, X. (2018). The impacts of rural tourism on the vulnerability of farmers' livelihood: From the perspective of coupled social-ecological systems. *Research of Agricultural Modernization*, 39(4), 654–664. doi:10.13872/j.1000-0275.2018.0040.

Chambers, R. & Conway, G. (1992). Sustainable rural livelihoods: Practical concepts for the 21st century. *Institute of Development Studies (UK)*.

Guo, Z.Y., Xie, S.Y., Qiao, H.F., Xu, J.B., & Jia, Y.Y. (2022). *Identification and assessment of the poverty-returning risks facing households in rural tourism sites: A case study of Enshi Tujia and Miao Autonomous Prefecture of Hubei Province.* Research of Agricultural Modernization. doi:10.13872/j.1000-0275.2022.0047.

Han, L., Qiao, H.F., & Xie, S.Y. (2019). The difference and classification of residents' perception on tourism impact in Enshi tourism poverty alleviation villages based on PCI. *Resources Science*, 41(2), 381–393.

Huddart, D. & Stott, T. (2019). *Outdoor Recreation: Environmental Impacts and Management*. Palgrave Macmillan.

Jia, Y. Y., Hu, J., Xie, S. Y., Qiao, H. F., & Liu, D. J. (2021). Vulnerability and influence mechanisms of socialecological system in poor mountainous tourism destination. *Human Geography*, 36(01), 155–164.

Lo, F., Gau, L., Hsu, Y., Gage, R., & Hung, C. (2021). An impact model for experiential activities of farming and food education in agricultural sustainable tourism. *e-Review of Tourism Research*, 18(4), 1–24.

Macfarlane, R. (2004). *Mountains of the mind: Adventures in reaching the summit*. Vintage Publishing.

Marek, A., & Wieczorek, M. (2015). Tourist traffic in the Aconcagua massif area. *Quaestiones Geographicae*, 34(3), 65–76.

Mayor, A. (2021). *Who were the first recreational mountain climbers?* Retrieved 29/11/2023 from https://www.wondersandmarvels.com/2013/06/who-were-the-first-recreational-mountain-climbers.html.

Messerli, B., & Ives, J. (1997). *Mountains of the world: A global priority*. Parthenon Publishing.

Neba, N.E. (2010). Developing rural tourism as an alternative strategy for poverty alleviation in protected areas: Example of Oku, *Cameroon. International NGO Journal*, 5(1), 50–59.

Nepal, S. K. (2000). Tourism in protected areas: The Nepalese Himalaya. *Annals of Tourism Research*, 27(3), 661–681. https://doi.org/10.1016/S0160-7383(99)00105-X.

Nepal, S.K. & Chipeniuk, R. (2005). Mountain tourism: Toward a conceptual framework. *Tourism Geographies*, 7(3), 313–333.

Nepal, S. K. (2008). Tourism-induced rural energy consumption in the Annapurna region of Nepal. *Tourism Management*, 29(1), 89–100. https://doi.org/10.1016/j.tourman.2007.03.024.

Nyaupane, G. P. (2015). Mountaineering on Mt Everest: Evolution, economy, ecology and ethics. In G. Musa, J. Higham, & A. T. Carr (Eds.), *Mountaineering tourism*. Routledge.

Qiao, H.F., Xie, S.Y. (2022). *Performance evolution of tourism poverty-alleviation policies and prevention of poverty-returning risks at the mountainous area in western Hubei.* China Social Science Press.

Statistics Bureau of Enshi Tujia and Miao Autonomous Prefecture. (2020). *2019 Statistical Communiqué on National Economic and Social Development of Enshi Prefecture.* https://www.stats.gov.cn/english/PressRelease/202002/t20200228_1728917.html.

Tao, T. & Wall, G. (2009). Tourism as a sustainable livelihood strategy. *Tourism Management*, 30(1), 90–98. https://doi.org/10.1016/j.tourman.2008.03.009.

Wang, H., Xiong, L. & Gage, R. (2021). Cultivating destination brand ambassadors in rural China: Examining the role of residents' welcoming nature. *International Journal of Tourism Research*, 23(6), 1027–1041. https://doi.org/10.1002/jtr.2460.

Wang, Q., Yin, M.H., Yang, X.Z. & Yao, Z. (2019). Spatio-temporal evolution and impact mechanism of socioecological system vulnerability in poor mountainous tourist destinations: Taking Dabie Mountain Area as example. *Acta Geographica Sinica*, 74(08), 1663–1679. https://doi.org/10.11821/dlxb201908013.

Xie, S.Y., Li, L., Feng, J. & Qiao, H.F. (2020). A study on the differences between the poor and non-poor residents' perception on tourism poverty alleviation policy performance in Enshi. *Tourism Tribune*, 35(02), 80–92.

Zhao, X.Y. (2017). Sustainable livelihoods research from the perspective of geography: The present status, questions, and priority areas. *Geographical Research*, 36(10), 1859–1872.

Zurick, D. N. (1992). Adventure travel and sustainable tourism in the peripheral economy of Nepal. *Annals of the Association of American Geographers*, 82(4), 608–628. https://doi.org/10.1111/j.1467-8306.1992.tb01720.x.

4 Inclusive Development as a Modern Need

Can Mountain Tourism Make a Difference in the Caribbean?

Renata Rettinger

Introduction: Tourism, Development and Pro-Poor Action

Tourism, as a branch of the economy, creates an array of benefits as well as negative outcomes (Mason, 2020). It serves as an economic saviour for many geographic regions – impoverished regions. Tourism helps to redistribute wealth from the relatively affluent North to the much poorer South (Buckley et al., 2015; Apollo & Moolio, 2021; Schänzel & Apollo, 2023). However, the economic benefits of tourism in the global South are a matter of debate (Cheer, 2020; Rogerson & Saarinen, 2018). Tourism is sometimes overvalued as a source of economic development in the Global South.

A prime example is islands with mountainous areas. Peripheral mountainous areas do not take over the dominant tourism function, while island tourism based on the 3S principle is likely to dominate (Apollo & Rettinger, 2019). This type of tourism is based on large hotel complexes offering all-inclusive tourist packages (Rettinger & Apollo, 2017). Foreign investors usually own these hotels. Many studies suggest that some negative side effects of this type of tourism can be observed, ranging from economic to social and cultural (Matthews, 1977). However, a well-developed tourism economy does not mean economic prosperity for all residents. Tourism can also cause negative side effects for the region where it is concentrated (Schänzel & Apollo, 2023). Poverty may increase, and some local social groups may be marginalised. In many countries, not all residents share the costs and benefits of tourism equally. The relationship between tourism development and overall local development in a given geographic region may also be analysed in terms of indirect benefits, as opposed to direct financial benefits, manifested by the so-called multiplier effect of tourism-related income in the local economy (Gałecki, 2004). If tourist expenditures are exported out of the country where the tourist region is located, then the multiplier effect will be negligible in that country. One of the critical research areas should be the relationship between growth centres and peripheral areas (Rettinger & Apollo, 2017).

A key issue in this chapter will be to examine the position of tourism in the regional economic system, analysed from the perspective of intersectoral linkages. Unfortunately, in many countries, tourism development occurs in unique enclaves, severely limiting or eliminating intersectoral linkages – the

DOI: 10.4324/9781003392583-5

concept of cathedrals in the desert. One of the reasons for this may be insufficient growth in other sectors of the economy, which are unable to meet the growing demand for goods and services in the tourism sector or are unable to provide an adequate level of quality. Consequently, the island regions of the Southern countries seek foreign investors, who then export most of their hard currency profits outside the destination country. This reduces the multiplier effects on the host country's tourists. In many island countries, the system appears to be concreted over. Activating mountain areas can be a solution and, in a way, show the way for the future development or redevelopment of coastal areas to bring more benefits to the local community.

In the Caribbean, both coastal and mountain regions can be found. While the Caribbean's seaside is well known due to all-inclusive mass tourism (3S), its mountains are rarely visited, if not forgotten. Even if some paper has highlighted their tremendous potential and the possibility of becoming a key to product diversification in tourism (Prosser, 1983; Nelson, 2011; Rettinger & Apollo, 2017; Weaver, 1991), to understand this, however, it is necessary to look at the Caribbean region holistically. For this reason, it was decided to trace the primary tourism model that the Caribbean regions are developing and to weave the mountain regions into it.

Caribbean, the All-Inclusive Dominant Tourist Region

The Caribbean is one of the world's most rapidly developing tourist regions. A variety of endogenous and exogenous factors determines its rate of growth. On the other hand, tourism also helps drive other economic development through foreign investment, new jobs, and the influx of hard currency to most Caribbean nations. The Caribbean destinations received an estimated 32.0 million international tourist arrivals in 2019, 4.4% or some 1.3 million more than the 30.7 million in 2018 (CTO, 2024). The Caribbean is one of the world's most tourism-dependent regions, with 12.8% of GDP generated by the tourist sector. This can be risky to the region's economy, as most of its tourists arrive from outside the region, and income from tourism strongly depends on the economic state of the countries sending tourists to the Caribbean.

The tourist sector is one of the largest employers in the Caribbean, directly or indirectly providing jobs for 2.1 million workers. In some Caribbean countries, 25% of the workforce is involved in tourism. Only the government sector is larger than the tourist sector in the region. The Caribbean tourist sector is estimated to employ 2.76 million workers by 2021. Tourism may help alleviate unemployment and poverty and accelerate development in rural areas. For this to happen, two things must occur first. Caribbean countries must create a business climate that is friendly to foreign investors and support local communities in their attempts to start local businesses. The role of tourism ought not to be limited to the influx of hard currency. The reason for this is straightforward – most hard currency flows in from foreign tourists and later flows into foreign investor bank accounts. Tourism should also help spark local development, which would help gradually transform the societies and economies of the Caribbean region.

Tourism in the Caribbean is often viewed purely in terms of large corporations, foreign investment, and government policy. The three areas of activity often appear to be the only viable alternatives for the region. The research literature is focused on the dominant model of tourism development in the Caribbean, which consists of all-inclusive resorts offering a standard package of attractions. Smaller tourist businesses, local governments, and local communities are treated as passive entities simply reacting to the activity of global corporations. What is interesting and practical in research is the role of local communities in the tourist sector in the Caribbean. What can be even more interesting is the view of economic development from the perspective of local communities. In this case, many research papers focus mainly on social and economic factors associated with tourism and omit cultural and environmental factors.

Many research papers do not question the significance of foreign investment in the development of the tourist industry. However, some papers on local development (Domański, 2001; Grabher, 1994;) are more critical of the role of foreign direct investment. Grabher (1994) describes it as establishing "cathedrals in the desert." This is the notion that foreign investment often creates exogenous enclaves without meaningful linkages to the local economy. The basis of the global capitalist system is the pursuit of ever-greater profits via acquiring new markets and opportunities to reduce production costs (Domański, 2001). In many tourist regions, this pursuit becomes somewhat pathological in that tourist towns become fenced enclaves.

The construction of large, all-inclusive hotel complexes in coastal areas has been the principal trend in the Caribbean tourist industry since the early 1990s. Both all-inclusive (Issa & Jayawardena, 2002) and cruise-based options are the mainstay of the Caribbean tourist sector. Both options rely on tourist enclaves. Research has shown that many foreign tourists select Caribbean destinations primarily based on the "enclave" factor. Other factors include the beautiful natural environment, non-material cultural heritage, all-inclusive tourist centres, luxury hotels, transportation options, safe beaches and hotels, as well as the region's "tropical climate" (Lee & King, 2006). Tourists staying at tourist enclaves do not have the opportunity to meet many residents and encounter local culture. Consequently, most tourists spend money at the hotel complex, which does not significantly benefit the local population and the host country.

Many research publications suggest that the Caribbean region's tourist offering needs to change. A variety of trends and problems in the tourist sector in Jamaica are discussed by Boxhill (2007). The paper emphasises the need to focus the tourist offering more on Jamaican culture and history and suggests that more residents should be involved in the tourist sector. The Caribbean is losing its position in the world tourist market to other destinations, including Latin America, Asia, Florida, and the South Pacific. While the

Caribbean is very competitive, it does not fully utilise its immense competitive potential. One reason for this is the region's focus on exclusive tourism based on high tourist expenditures per capita (Theuns, 2011).

In light of the above, the research will focus on identifying endogenous factors of business development in the tourist sector and growth centres based on endogenous factors. Given their dependence on local factors, foreign capital is less often invested in peripheral regions. Local aspirations, innovativeness, attitudes towards work, and models of social behaviour can all contribute to permanent social and economic success. In addition, this area of research will include an analysis of success factors in the context of path dependence analysis.

The factors driving tourism development – especially leisure tourism – in tourist regions are virtually identical – Sea, Sand, and Sun (3S). Both old and new tourist regions must compete with one another in an attempt to attract new tourists. Small but very competitive islands will win. This is the reason why product differentiation is a critical strategic goal. Research on hotel services has shown that human capital can make a crucial difference in a competitive marketplace. One key differentiating factor associated with human capital is cultural heritage – especially non-material cultural heritage (local traditions and customs). The residents of the Caribbean region are characterised by a vibrant and diverse culture, which, combined with a beautiful natural environment, can serve as a strong magnet for new tourist traffic.

Furthermore, this culture and unspoilt landscape is best preserved outside the coastal areas, especially in the much less (or not at all) transformed inland areas (i.e. mountains). Likewise, the tourist sector can operate year-round due to its warm climate and tourist flow from several directions. Moreover, the central location of the Caribbean is also beneficial, as it is easily accessible to tourists from the United States, Canada, and the upwardly mobile region of South America.

It is also important to mention that the trend of motivations that tourists follow is changing – and while all-inclusive mass tourism (3S) will most likely always dominate in the Caribbean area, openness to new elements (e.g. adventure, active, cultural tourism) exists and will become increasingly popular (Rettinger & Apollo, 2017). Studies from other areas confirm this trend and indicate precisely the mountains' potential to achieve well-being (Ritpanitchajchaval, Ashton & Apollo, 2023). Moreover, mass tourists are also striving for well-being. Thus, mountain tourism can be a way not only to activate peripheral areas but also to change the all-inclusive model to one that is more inclusive.

Inclusive Tourism as a Development Opportunity for the Caribbean Region

Inclusiveness is the reconciliation of social, economic and environmental objectives while optimising the use of socio-economic potential (Scheyvens & Biddulph, 2018). In the tourism economy, the idea of inclusive development is based on widening the range of people participating in and benefiting from tourism activities. One of the most critical assumptions is that tourism reduces poverty and drives inclusive development. Unfortunately, in the case of many countries in the Global South, governments are interested in developing international tourism, which is unfortunately very often limited to all-inclusive tourism only. This type of tourism generates higher income, and its flow is controllable by the country's tax system. Afterwards, the distribution of profits is very unfair and even often restricted to a narrow group in power. Of course, according to macroeconomic indicators, economic growth is recorded in such countries, and tourism contributes substantially to this. Unfortunately, this type of tourism causes social inequality.

In many countries or regions, recent years have seen an intensive development of the concept of growth through the tourism economy. This growth is understood to be creating new jobs, increasing budget revenues, and preserving natural and cultural heritage. The increasing number of tourists worldwide can lead to processes of degradation of the tourist space (loss of tourist assets), i.e. "in the glare of tourism, you can get burned" (Biddulph & Scheyvens, 2018). Therefore, initiatives should be taken to prevent such phenomena and mitigate the effects of the "unwise development" of tourism (Scheyvens, 2006).

The dispersion and expansion of the areas of activity of the tourism industry and its anchoring in all sectors of the economy predisposes it to participate in inclusive development systemically. Here, too, the untapped potential of mountain areas undeniably emerges. Thus, it may become an opportunity to accelerate the reduction of civilisational backwardness, resulting in comprehensive socio-economic growth in the context of the sustainability of these isolated areas – e.g. mountains. They can become the beginning of a new opening for greater public involvement and serve as a lesson for modernising and changing the approach to coastal areas. Moving (even slowly) away from "unwise developed" tourism is the challenge facing tourism today.

Inclusive Development: A Present-Day Demand

Economic inclusiveness implies that GDP growth is not an end in itself; it is more important to distribute this growth's benefits and redress income and wealth inequalities. This is essential for creating the conditions for harmonious socio-economic development. Such development is meant for inclusive development. Inclusiveness, in turn, is understood here as the reconciliation of social, economic and environmental goals while optimising the use of socio-economic potential. Thus, economic growth (i.e. GDP growth) is not seen as a single measurement but as one of three indicators, including social and environmental progress, as pillars of socio-economic development.

Different publications have offered various definitions of inclusive growth. One such definition describes it as transformational tourism, a concept that involves the engagement of marginalised groups in the ethical production or consumption of tourism, ensuring that they share in its benefits (Biddulph, Scheyvens, 2018). As defined by the World Bank, the hallmark of inclusive growth is a focus on both the pace and structure of growth. Rapid and sustained poverty reduction requires inclusive growth through which communities contribute to and benefit from economic growth. Rapid growth rates are undoubtedly necessary for significant poverty reduction, but for growth to be sustainable in the long term, it should be broad-based across sectors and encompass a large proportion of a country's or region's workforce. (World Bank, 2009).

The inclusive growth paradigm falls within the framework of sustainable development and can be seen as a process or policy response (Hasmath, 2015) as well as a "desired outcome" (George, McGahan & Prabhu, 2012). Inclusive development should lead to economic growth that is felt in both low- and middle-income households (de Haan & Thorat, 2013). The potential of this approach is clearly demonstrated by research in mountainous regions such as the Himalayas (Wengel, 2021; Apollo & Andreychouk, 2022) and the Caribbean (Apollo & Rettinger, 2019 Srinivasan et al., 2017; George, 2020), both of which are economically and politically marginalised areas. This underscores the significant gains that isolated mountain areas stand to achieve through such an inclusive growth approach.

Tourism economy in the context of inclusive development

The inclusiveness of tourism can be seen as an ambition of global policy and grassroots initiatives. The modern tourism economy is characterised by heterogeneity, in which there is an interconnection and interaction of various industries such as transportation (international and domestic), accommodation, catering, guiding and trade. This dispersion of the tourism industry and even its anchoring in all divisions of the national economy predisposes it in a systemic way to participate in inclusive development (Hampton, Jeyacheya, & Long, 2018).

Tourism in the Global South is driving the emergence of both formal and informal sectors, notably fostering the growth of the informal sector and promoting social inclusion (Kozel, 2014). Many studies emphasise the need for development in the long term, but at the same time, the question is formulated as to whether tourism-driven growth will become possible as a short-term or long-term development (Hampton, Jeyacheya, & Long, 2018). However, it should be mentioned that in the Global South, tourism-related development opportunities are impossible among the poor and marginalised (Apollo & Moolio, 2021; Harrison, 1992; Jamal & Camargo, 2014). This is particularly true in peripheral areas, where change is virtually impossible without institutional assistance.

Inclusive tourism development should lead to more equitable and sustainable results. Inclusivity in the tourism economy includes a variety of actors in different environments, seeking to expand the range of people involved in creating and consuming tourism products. Inclusive tourism development seeks not only to expand access to consumption, production and benefit sharing in existing tourism destinations but also to redraw the tourism map to create new places of experience and interaction (Biddulph & Scheyvens, 2018). Such a "map" is illustrated in the following two sections.

All-inclusive 3S: an exclusionary trend in tourism

The question should be asked whether specialisation excludes diversification and vice versa (Sheng, 2011). In the case of the Caribbean, there is a regularity that the determinants of tourism development in most countries are the 3S principle (sand, sea, sun), while of great importance are the diversifying elements such as natural (including marginalised mountain areas) or anthropogenic qualities that allow the development of other forms of tourism.

The biggest criticism is the exclusive nature of tourism. This exclusivity means the creation of tourist enclaves and the transfer of profits to large companies, very often based abroad. All-inclusive tourism is a heavily criticised trend in modern tourism, but unfortunately, it has numerous supporters from both the demand and supply sides. Tourists choose to stay in all-inclusive resorts because they offer more opportunities for planning and enjoying their relaxation (Buhalis, 2000), the exact cost of the proposed rest offer, and the issue of safety in remote, "exotic" destinations (Wong, Kwong, 2004). Safety in travel can also be considered as sanitary or cultural safety, so rest in resorts provides a certain level of service (Enoch, 1996). The benefits of this form of tourism for tour operators include easier sales of service packages, focus on hotels, and product repeatability based on the service package. The weaknesses of all-inclusive tourism include its failure to generate expenses outside the hotel and its restriction of interactions with local residents. In the literature, one may encounter strong phrases such as "entrapment" within the hotel (Issa & Jayawardena, 2002. The aspect of elitism of leisure also refers to the exclusivity of the hotel, the standard of the accommodation in its location and the food and beverage offered, which are of enormous importance (Chu, Choi, 2000; Taninecz, 1990; McCleary, Weaver & Hutchinson, 1993; Anderson et al. 2009).

Can the development of all-inclusive tourism in the Global South be considered a kind of systemic dysfunction, and is the implementation of the assumptions of inclusive tourism an unfeasible project? To some extent, this hypothesis is true. Contrary to appearances, the strong primacy of all-inclusive tourism and the high degree of spatial concentration of development and tourist traffic can create an excellent climate for alternative activities that escape a particular development pattern. Implementing new solutions in a country perceived as an all-inclusive tourism region is not easy, which does not mean it is impossible. Such is the case in many tourist regions where steps have been taken to create a new socio-economic order.

The previous model of tourism development in the Caribbean, which favours tourist enclaves, i.e. high-category tourist resorts and roundtrip cruises, has led to the exhaustion of tourism from elements indicative of the areas' past and present culture. This phenomenon is incompatible with contemporary trends in world tourism, and where there is a shift from 3S to 3E (education, excitement, entertainment) passive type tourism, i.e., there is an apparent domination of cultural tourism over tourism based on natural assets. The continued creation of tourist enclaves with highly categorised all-inclusive hotels contradicts global tourism trends, which are based on intangible attractions and appeals to feelings, experiences, impressions, social and psychological climate. All these transformations in tourism are primarily due to changes among tourists. The modern tourist is more affluent, more demanding, better educated, has a broad worldview, and seeks contact with the wider reception environment. Changing the model of tourism development in the Caribbean will benefit potential tourists and, more importantly, create opportunities for local communities. This change will set in motion the processes of activating endogenous potential, which is relatively underutilised in the all-inclusive resort tourism model (Rettinger & Apollo, 2017). Mountain areas, with their magic and wildness but also authenticity – far superior to that of the coasts – can drive this positive change.

Caribbean Mountain tourism: an inclusive trend in tourism

The islands of the Greater Antilles archipelago (Jamaica, Cuba, Haiti, Puerto Rico) belong to a group of islands with very good tourism development for leisure tourists. However, the potential lies in the central parts of the islands, i.e., the mountainous areas. All the islands mentioned above exhibit a notable spatial polarisation, with well-developed coastal regions designated as tourist centres for leisure activities, contrasted against the underdeveloped central mountainous areas, which are peripheral compared to the coastal regions and are scarcely developed for tourism.

Numerous reasons contribute to this situation, with the most significant ones including the motives of foreign tourists seeking 3S coastal recreation, along with inadequate tourist infrastructure (accommodation, dining, and supporting facilities), limited transportation accessibility, and insufficient promotion of the Careian mountain areas as tourist destinations by tour operators (Apollo & Rettinger, 2019; Rettinger & Apollo, 2017). Frequently, mountain regions serve as tourist reception areas for optional tours, often termed "sightseeing on occasion." More significantly, these mountain regions are regarded as locations of significant cultural authenticity, enhancing their appeal. Alternatively, they may also be perceived as impoverished areas, which can intrigue tourists from affluent nations.

When examining tourism development in mountain regions within a tourism context, a crucial question arises: what factors contribute to the lack of dynamic tourism growth in these areas? One primary reason is the prevalent association of islands with seaside vacations, often organised within all-inclusive resorts providing comprehensive services. Moreover, travel agencies and tour operators often instil in tourists a perception of external dangers (such as theft, robbery, and extortion), compelling them to seclude themselves from the local population. This prompts further inquiry into the supply side of this phenomenon, mainly whether mountain residents are open to and prepared to engage with tourism development to secure its benefits. Experience from several regions in the Dominican Republic and Jamaica indicates real opportunities for tourism development in mountain areas based on the endogenous potential inherent in the local population.

Nevertheless, unfortunately, several barriers fundamentally limit the development of tourism entrepreneurship in mountain areas. An essential factor conditioning the formation of entrepreneurial attitudes is the need for initial capital to start a business. Unfortunately, in the case of the Caribbean islands, this is the primary problem: the lack of financial resources for investment, whether it is the creation of accommodations, food and beverage outlets or elements of the accompanying base. The constrained financial capabilities stem from the poverty experienced by the population residing in mountainous regions, many of whom lack permanent employment opportunities. Residents of countries such as Cuba, the Dominican Republic and Jamaica receive funds from abroad through direct transfers from families living most often in the US. Unfortunately, studies indicate that these direct transfers are used for current consumption and are rarely used for investment purposes. Some exceptions to this rule are Cuban citizens, who are allowed to start *casas particulares* as a form of accommodation base, i.e. guest rooms. On the territory of Cuba, such facilities are plentiful, and they are a crucial element that complements the country's tourist development (Rettinger, 2014). In many countries, it is necessary to regulate the conditions for the operation of businesses. An essential element of economic activation of mountain regions is preparing an attractive tourist offer, in which the core of the tourist product will be mountain areas. Based on endogenous potential, i.e. elements of the natural environment and cultural heritage of the local population, it is possible to create a unique tourist product, different from recreation in coastal areas in chain all-inclusive hotels. A prerequisite will be to determine the target group for such recreation, i.e. we must ask ourselves who will come and for how long to relax in the mountain areas of the islands. It is necessary to conduct a detailed analysis of development prospects in the context of complementary offerings, i.e. facilities for optional excursions, which will show another side of tourism on the islands. An essential aspect of tourism development in the islands' mountain areas is the need to introduce the principles of sustainable development.

Another important question to consider is whether there are viable tourism development opportunities in island nations' mountainous areas. The answer is not just affirmative but imperative. This is substantiated by domestic tourism, as residents from diverse regions frequently seek relaxation within short distances from their homes, often opting for weekend getaways. In the Caribbean, residents, constrained by financial limitations, frequently find respite in the central areas of the islands, along riverbanks or amidst mountainous surroundings, rather than indulging in costly coastal hotel stays. An excellent example of this phenomenon is two regions in the Dominican Republic, namely Rio Partido in Dejabon province and Damajagua Falls. These two regions owe their success to domestic tourism and their location in a mountainous region. There are many such little-known places (for foreign tourists) that require investment for tourism development. Due to the history of economic leakages in tourism and hospitality in other locations in the Caribbean region, these mountain locations lack interest in foreign investors. Thus, Caribbean mountain locations highlight a promising development opportunity for the local population. Presently, and in the future, these areas will continue to serve as recreational hubs for the local populace, thereby enhancing the inclusive nature of the tourism economy and fostering opportunities for local involvement in tourism, particularly in domestic tourism.

Conclusion and Future Research

The era of mass tourism in the Caribbean took root in the mid-20th century. Presently, the Caribbean is popularly associated with the 3S (Sun, Sand, and Sea) and, for some, the 3E (Education, Excitement, Entertainment). However, with heightened competition and evolving tourist motivations, regional authorities are compelled to diversify their offerings beyond the traditional 3S and 3E paradigms. Consequently, Caribbean authorities actively seek out new categories of tourists, such as those inclined towards more adventurous experiences. Appealing pristine, isolated mountainous areas of the Caribbean region hold great potential for pro-poor mountain tourism development. It is important to emphasise that mountain tourism has long been regarded as a driver of economic and social progress, particularly in mountainous regions within economically disadvantaged, traditional, and developing countries (Apollo & Wengel, 2022).

The mountains of the Greater Antilles, including those in Cuba, Hispaniola, Puerto Rico, and Jamaica, undoubtedly present an opportunity for mountain tourism development, which may significantly contribute to poverty alleviation, particularly for communities inland. However, making these mountain areas accessible to tourists is critical for developing a viable local tourism industry (Apollo & Rettinger, 2019). The level of mountain tourism accessibility is critical to improving destination accessibility in those areas. These improvements can potentially contribute to mountain regions' economic and social development by attracting tourists when they choose their destination (Apollo, 2017).

A fundamental problem is the tourism development model in the Caribbean, where two development trends prevail: leisure tourism and circumnavigation. In a way, this fundamentally limits the possibilities for entrepreneurial activities in tourism services. A detailed analysis of the status and development prospects of tourism in island resort regions forces us to reorient our research directions in the broad field of the tourism economy. Opening up to new areas can be an excellent opportunity for the island. The desired outcome should be the modification of the direction of tourist areas of the Caribbean region in terms of sustainable development and local community development. An attempt should be made to create a tourism model for the Caribbean region in light of the principles of sustainable development with a particular focus on stable community development. Future research should also focus on opportunities for sustainable tourism and so-called pro-poor tourism as novel means of improving economic conditions in low-income peripheral areas. Pro-poor tourism is not a new form of tourism but constitutes a new approach to tourism as a strategic route to reducing poverty in marginalised communities.

Management Implications

The main challenge facing tourism managers in the Caribbean is the emerging paradigm of the need to balance mass tourism with alternative tourism. The critical analysis of systemic dysfunctions carried out in this chapter, including the one-way development of the tourism region, clearly indicates not only the lack of pro-poor impact but even the expansion of poverty. Activation of mountain areas could change this situation. Of course, change will not happen immediately, but an all-inclusive policy can be supplemented with solution designs that, even if they cannot be implemented entirely at a given time, provide direction for recommendations directed toward economic policy (Miszewski, 2016). This approach also fosters the identification of barriers that – either permanently or temporarily – prevent the implementation of particular "utopian" projects. The results of such analyses can be helpful, even if some of the obstacles identified through them are deemed insurmountable. They can become the basis for defining long-term strategic goals.

Reference

Anderson, W., Juaneda, C., & Sastre, F. (2009). Influences of pro-all-inclusive travel decisions. *Tourism Review*, 64(2), 4–18. https://doi.org/10.1108/16605370910963482.

Apollo, M. (2017). The true accessibility of mountaineering: The case of the High Himalaya. *Journal of Outdoor Recreation and Tourism*, 17, 29–43. https://doi.org/10.1016/j.jort.2016.12.001.

Apollo, M., & Andreychouk, V. (2022). *Mountaineering Adventure Tourism and Local Communities: Social, Environmental and Economics Interactions.* Edward Elgar Publishing. https://doi.org/10.4337/9781802209389.

Apollo, M., & Moolio, P. (Eds.). (2021). *Poverty and Development: Problems and Prospects.* Channel View Publications. https://doi.org/10.21832/9781845418489.

Apollo, M., & Rettinger, R. (2019). Mountaineering in Cuba: Improvement of true accessibility as an opportunity for regional development of communities outside the tourism enclaves. *Current Issues in Tourism*, 22(15), 1797–1804. https://doi.org/10.1080/13683500.2018.1446920.

Apollo, M., & Wengel, Y. (2022). *Mountaineering tourism: A critical perspective.* Routledge. https://doi.org/10.4324/9781003095323.

Boxhill, I. (2007). *The Political Culture of Democracy in Jamaica, 2006*. LAPOP.

Buckley, R., Gretzel, U., Scott, D., Weaver, D., & Becken, S. (2015). Tourism mega-trends. *Tourism Recreation Research*, 40(1), 59–70. https://doi.org/10.1080/02508281.2015.1005942.

Buhalis, D. (2000). Relationships in the distribution channel of tourism: Conflicts between hoteliers and tour operators in the Mediterranean region. *International Journal of Hospitality & Tourism Administration*, 1(1), 113–139. https://doi.org/10.1300/J149v01n01_07.

Cheer, J. (2020). Tourism and Its Discontents in the Global South. *Current History*, 119(813), 29–33. https://doi.org/10.1525/curh.2020.119.813.29.

Chu, R. K., & Choi, T. (2000). An importance-performance analysis of hotel selection factors in the Hong Kong hotel industry: a comparison of business and leisure travellers. *Tourism management*, 21(4), 363–377. https://doi.org/10.1016/S0261-5177(99)00070-9.

CTO (2019). *CTO Annual Statistical Report*. Retrieved: https://www.onecaribbean.org/buy-cto-tourism-statistics/annual-statistical-report/.

De Haan, A., & Thorat, S. (2013). *Inclusive growth: More than safety nets.* International Development Research Centre, Ottawa.

Domański, B. (2001). *Kapitał zagraniczny w przemyśle Polski: prawidłowości rozmieszczenia, uwarunkowania i skutki.* Kraków: Instytut Geografii i Gospodarki Przestrzennej UJ.

Enoch, Y. (1996). Contents of tour packages: A cross-cultural comparison. *Annals of Tourism Research*, 23(3), 599–616. https://doi.org/10.1016/0160-7383(96)00001-1.

Grabher, G. (1994). The disembedded regional economy: the transformation of East German industrial complexes into western enclaves. In Amin, A. and Thrift, N. (Eds), *Globalization, Institutions and Regional Development* in Europe (177–195.). Oxford University Press.

Gałecki, R. (2004). Znaczenie efektu mnożnikowego wpływów z turystyki dla rozwoju regionu i miejscowości In Gałecki, R., (Ed.). *Rola i zadania samorządu terytorialnego w rozwoju gospodarki turystycznej* (113–117.). Mazowiecka Wyższa Szkoła Humanistyczno-Pedagogiczna w Łowiczu.

George, B. (2020). Inclusive sustainable development in the Caribbean region: Social capital and the creation of competitive advantage in tourism networks. *Business Ethics and Leadership* 4(3), 119–126. DOI:10.21272/bel.4(3).119-126.2020.

George, G., McGahan, A. M., & Prabhu, J. (2012). Innovation for inclusive growth: Towards a theoretical framework and a research agenda. *Journal of Management Studies*, 49(4), 661–683. https://doi.org/10.1111/j.1467-6486.2012.01048.x.

Hampton, M. P., Jeyacheya, J., & Long, P. H. (2018). Can tourism promote inclusive growth? Supply chains, ownership and employment in Ha Long Bay, Vietnam. *The Journal of Development Studies*, 54(2), 359–376. https://doi.org/10.1080/00220388.2017.1296572.

Hardy, J. (1998). Cathedrals in the desert? Transnationals, corporate strategy and locality in Wrocław. *Regional Studies*, 32(7), 639–652.

Harrison, D. (1992). International tourism and the less developed countries: The social consequences. In D. Harrison (Ed.) *Tourism and the Less Developed*. Chichester: Belhaven Press, 19–34.

Hasmath, R. (2015). The paradigms of inclusive growth, development and welfare policy. In R. Hasmath (Ed.), *Inclusive Growth, Development and Welfare Policy* (13–21). Routledge.

Kozel, V. (Ed.). (2014). *Well begun but not yet done: Progress and emerging challenges for poverty reduction in Vietnam*. World Bank Publications.

Issa, J., & Jayawardena, C. (2002). *All-inclusive business in the Caribbean. Caribbean Tourism: People, Service and Hospitality*. Ian Randle Publisher..

Jamal, T., & Camargo, B. A. (2014). Sustainable tourism, justice and an ethic of care: Toward the just destination. *Journal of Sustainable Tourism*, 22(1), 11–30. https://doi.org/10.1080/09669582.2013.786084.

Lee, C. F., & King, B. (2006). Assessing destination competitiveness: An application to the hot springs tourism sector. *Tourism and Hospitality Planning & Development*, 3(3), 179–197.

Matthews, H. G. (1977). Radicals and third world tourism: A Caribbean focus. *Annals of Tourism Research*, 5, 20–29.

Mason, P. (2020). *Tourism impacts, planning and management*. Routledge.

McCleary, K. W., Weaver, P. A., & Hutchinson, J. C. (1993). Hotel selection factors as they relate to business travel situations. *Journal of Travel Research*, 32(2), 42–48.

Miszewski, M. (2016). Inkluzywny system gospodarczy–utopia i droga do jej realizacji. *Studia i Prace WNEiZ US*, 44(2), 237–254.

Nelson, V. (2011). The landscape reputation: Tourism and identity in the Caribbean. *Tijdschrift voor economische en sociale geografie*, 102(2), 176–187. https://doi.org/10.1111/j.1467-9663.2010.00608.x.

Prosser, R. (1983). Caribbean tourism. *Journal of Geography*, 82(5), 243–249. https://doi.org/10.1080/00221348308980407.

Rettinger, R. (2014). Polityka lokalizacyjna wybranych systemów hotelowych na Kubie w zmieniających się realiach politycznych. *Prace Komisji Geografii Przemysłu Polskiego Towarzystwa Geograficznego*, (28), 9–22.

Rettinger, R., & Apollo, M. (2017). Enriching tour-operators' offer to go beyond sea, sun and sand: The case of the Dominican Republic. In R. Efe and M. Öztürk (Eds),*Contemporary Studies in Environment and Tourism* (60–81). Cambridge Scholars Publishing

Ritpanitchajchaval, N., Ashton, A. S., & Apollo, M. (2023). Eudaimonic well-being development: Motives driving mountain-based adventure tourism. *Journal of Outdoor Recreation and Tourism*, 42, 100607. https://doi.org/10.1016/j.jort.2023.100607

Rogerson, C. M., & Saarinen, J. (2018). Tourism for poverty alleviation: Issues and debates in the global South. In Chris Cooper, S. Volo, W.C. Gartner and N. Scott (Eds.), *The SAGE handbook of tourism management: Applications of theories and concepts to tourism* (22–37). Sage.

Schänzel, H., & Apollo, M. (2023). Specialty grand challenge: wherever tourism arrives, society notes it first!. *Frontiers in Sustainable Tourism*, 2, 1341765. https://doi.org/10.3389/frsut.2023.1341765.

Scheyvens, R., & Biddulph, R. (2018). Inclusive tourism development. *Tourism Geographies*, 20(4), 589–609. https://doi.org/10.1080/14616688.2017.1381985.

Sheng, L. (2011). Regional competition and sustainable development: a game theory model for tourism destinations. *European Planning Studies*, 19(4), 669–681.

Srinivasan, M. K., Otker, M. I., Ramakrishnan, M. U., & Alleyne, M. T. S. C. (2017). *Unleashing Growth and Strengthening Resilience in the Caribbean*. International Monetary Fund.

Taninecz, G. (1990). Business traveller survey. *Hotel and Motel Management*, 57(1), 29–32.

Theuns, H. L. (2011). A structural review of the extent of actual and potential leakages/linkages in Caribbean tourism. *Folia Turistica*, 25(1), 115–144

Weaver, D. B. (1991). Alternative to mass tourism in Dominica. *Annals of tourism research*, 18(3), 414–432.

Wengel, Y. (2021). The micro-trends of emerging adventure tourism activities in Nepal. *Journal of Tourism Futures*, 7(2), 209–215. https://doi.org/10.1108/JTF-01-2020-0011.

Wong, C. K. S., & Kwong, W. Y. Y. (2004). Outbound tourists' selection criteria for choosing all-inclusive package tours. *Tourism Management*, 25(5), 581–592. https://doi.org/10.1016/j.tourman.2003.06.002.

5 Resident Quality of Life and a Wellness Community-Based Tourism Development Model

A Way of Poverty Eradication

Ann Suwaree Ashton

Introduction

This chapter aims to provide insights into residents' quality of life and a wellness community-based tourism (CBT) development model as a way of poverty eradication. As poverty and development are very complex phenomena, relevant stakeholders should aim to eradicate poverty by all means, and using tourism is one of them (Wani, Batool, Dada, & Shah, 2024). Many types of tourism can help eradicate poverty, especially within remote or local areas. For example, the development of mountain tourism and community-based tourism have tangible impact on poverty reduction (Apollo & Andreychouk, 2022; Apollo & Moolio, 2021; Gohori & van der Merwe, 2024).

Especially in this instance does CBT play an important role to increase the net benefits for the poor by ensuring the growth contributes to poverty reduction, which in turn leads to a significant change in community quality of life (Goodwin & Santilli, 2009; Harrison & Schipani, 2007; Wani et al., 2024). Quality of life (QoL) refers to a good life, and it is believed to be the same as living a high quality life (Ventegodt et al., 2003).

The development of community based tourism (CBT) destination has an impact via changes in the livelihoods of residents on economic, and environmental values, with a view for a better sense of well-being (Delita, Berutu, Tanjung, & Rambe, 2024; Nkemngu, 2015; Nur et al., 2021; Yunikawati et al., 2021). Nowadays, community based tourism (CBT) developments have become popular destinations for tourists who want to get a real authentic experience, a sensation of life with local culture and lifestyle (Curcija, Breakey, & Driml, 2019; López-Guzmán, Sánchez-Cañizares, & Pavón, 2011; Yusof, Ibrahim, Muda, & Amin, 2012). However, past research has shown that many previous studies focused heavily on CBT and resident attitudes, but failing to link how it impacts a resident's quality of life, and how to create community well-being (Khazaeipul & Ramezanzadeh Lasboyee, 2023; Makandwa, Makudza, & Muparangi, 2023; Nugroho & Numata, 2021; Yeap & Liow, 2024). Examples of past studies: an investigation on sustainability, and a conceptual framework on strategy (Yeap & Liow, 2024); women participating in CBT during

DOI: 10.4324/9781003392583-6

COVID-19 (Makandwa et al., 2023); heritage conservation and CBT (Qu, Zhang, Shen, & Olsen, 2023); the changes in resident attitudes toward CBT through destination development (Nugroho & Numata, 2021). Hence, past literature shows that quality of life and wellness CBT is rarely investigated especially in a Thailand context. In a rare study on resident attitude toward quality of life focused on the impact of CBT with the nearby community and residents (Blunsdon & Davern, 2007).

Significantly, it is important to develop quality of life strategy model, because first, tourism is a tool for poverty alleviation, and pro-poor tourism is an important implication for sustainability and enhancement of local resident's well-being; therefore CBT is one of the strategies that can enhance economic benefit that can help the locals (Chok, Macbeth, & Warren, 2007; Oei & McAlinden, 2014). The quality-of-life strategy in community based tourism (CBT) development can improve a resident's quality of life if managed and developed properly (Curcija et al., 2019; Kim, Uysal, & Sirgy, 2013; Uysal, Sirgy, Woo, & Kim, 2016). Particularly, CBT is a significant strategy to improve the situation of local communities by promoting greater economic empowerment and social benefits to all relevant stakeholders including residents, community, tourism and hospitality businesses (Goodwin & Santilli, 2009; Sebele, 2010). Finally, it is believed that resident quality of life makes for satisfaction and happiness, with an emphasis on economic responsibility through both material and spiritual well-being, social responsibility through community well-being, cultural responsibility through emotional well-being, and environment responsibility through health and safety well-being (Marans & Stimson, 2011; Mathew, 2022; Norman & Pokorny, 2017).

As discussed, this study provides insight into residents' quality of life and a wellness CBT development, as it is still lacking in both the academic and managerial perspectives. The main aim is to provide insight into resident quality life and a wellness community based tourism (CBT) development model as a way of poverty eradication. To achieve the study aim, three objectives were raised: first, to study resident subjective quality of life toward CBT development; second, to study resident objective quality of life toward CBT development; and third, to study resident existential quality of life toward CBT development.

Community Based Tourism and Resident Quality of Life

Quality of life (QoL) is an individual's perception of their position in life, usually within the context of culture and value systems, where it is mostly related to people's goals, expectations, standards and health concerns (Nadee & Ashton, 2021). It is a descriptive term that refers to people's emotional, social and physical well-being, their ability for functional, social and physical well-being, and their ability to function in the ordinary tasks of living (Donald, 1998, p. 5). It is a broad-ranging concept affected in a complex way by the person's physical health, psychological state, personal beliefs, social relationships and their relationship to the salient features of their environment (Ali, Quaddus, Rabbanee, & Shanka, 2022).

In relation to CBT, it is about tourism getting involved with sustainability, local community participation, and maximizing benefits to the community, and help the poor to have a better living. It is believed that there is a relationship between QoL and community based tourism; if residents have good QoL it means they have good health and education, economic improvement, good welfare, and live in a good environment, thus, these aspects are a result of a well-being community (Estes & Sirgy, 2019; Javed, Javed, & Khan, 2016).

Community well-being impacts resident quality of life and refers to resident attitude toward happiness and satisfaction in life, for instance, to feel happy and be part of the community based tourism development. The development of CBT must also have set goals for planning, implementing, and evaluating properly (Mathew & Sreejesh, 2017; Okazaki, 2008; Pinel, 1999). The key attributes of a community's wellbeing comprise, firstly, of social relations; it is one important domain that is needed within the community. It includes social networks, social support, social inclusion and exclusion, social cohesion, social capital, social justice, a sense of belonging, a sense of solidarity, respect and tolerance for diversity, gender equality, trust, reciprocity, security and safety, collaborative activities, local participation, and political participation (Raphael et al., 2001).

In summary, the quality of life concept for this study is referred to as a good life by those who believe it is same as living a life with high quality (Ventegodt et al., 2003). A high-level of well-being will be involved; the first direction in progress forward and upward towards a higher potential of functioning. Second is an open-ended and ever-expanding tomorrow with its challenges to live at a fuller potential. Third is the integration of the whole being of the total individual body, mind, and spirit in the functioning process (Dunn, 1959). Consequently, such sustainable development goals, especially the strategies helping the poor to have better opportunities in life, lead to improved income and education for future careers. Finally, life satisfaction helps with the sustainability triple bottom line: economic, socio-cultural, and the environment.

Quality of Life Theory

Much theory has been proposed on the quality of life (QoL), and it is often referred to as well-being; it is the objective of life conditions that apply to a population generally (Theofilou, 1995). QoL can be related to social approach, an example of this is such as Land and Spilerman's (1975) grouping a social indicator approach into three domains: objective conditions, subjective value-context, and subjective well-being.

First, objective conditions of life means how one's life is perceived by the outside world, and is influenced by the culture or external factors in which people live (Ventegodt et al., 2003). It can measured through the external physical and social conditions of an individual's existence, and this was used by FDHA (2016) to measure social conditions such as the setting of the environment where they live, such as crime rate, income, education, health access and social participation (Carley, 1979; Gregoire, 2002; Marans &

Stimson, 2011). In addition, it is measured based on jobs, housing conditions, health, environmental quality, personal security, civic engagement, work-life balance, infrastructure and services, mobility (a good transportation network is important, such as shopping, education and recreation), and culture & leisure (social interaction, clubs or sport facilities) (FDHA, 2016; Marans & Stimson, 2011).

Subjective value-context refers to an individual's beliefs, expectations, aspirations, and distributive justice value. It is about individual contentment and happiness with life, both aspects reflecting the subjective quality of life. Finally, subjective well-being refers to an individual's feelings, life satisfaction, specific satisfaction, or frustrations (derived from the first two domains) (Land & Spilerman, 1975). Thus, subjective well-being quality of life can be measured by a resident's overall sense of well-being based on life satisfaction, perceived quality of life, life domain satisfaction, for example, satisfaction in leisure life, social life, family life and work life, positive/negative affects, and overall happiness.

Another idea on the quality of life theory (Ventegodt et al., 2003) divides it into three separate groups, each concerned with an aspect of a good life. First, the subjective quality of life; second, the objective quality of life; and third, the existential quality of life. First, the subjective quality of life is about how each individual feels, but more importantly understands the meaning of life, happiness, and satisfaction where it leads to well-being and overall happiness in life (Sirgy, 2010).

Second, the objective quality of life is related to the external factors of life; it is about the fulfillment of people's life needs as perceived by the outside world, and it is influenced by cultural norms, especially where people live. It is about the ability of each individual adapting to the values of a culture; these can include social status or the status symbols that one should have to be a good member of that culture (Jeon, Kang, & Desmarais, 2016; Ventegodt et al., 2003). Hence, the residents' quality of life can be measured based on an aggregation of resident perception of economic, social and environmental conditions (Jeon et al., 2016).

Third, existential quality of life is related to what people believe about life and reality, and in the state of realization of life needs (Ventegodt et al., 2003, p. 1035). It can be described as spiritual involvement as all humans are seeking for their own self-actualization, such as the meaning of life, illness, death and a belief in a higher being (Davison & Jhangri, 2010). Therefore, spirituality is defined as more basic than religiosity, and is related to transcendent values and the way in which people find a purpose in life (Chan, 2018). Another meaning of the existential quality of life is how one's life is at a deeper level, especially how individuals deserve to be respected and live in a harmonious society or environment, including how one can have a healthy body, mind, and soul (Davison & Jhangri, 2010). Individual respect is about acknowledging that you know I am I, you are you, and the other is the other, you own your problems, as I own mine, everyone is intelligent and has his own resources to find his own solutions to his problems, hence it is about acknowledgement and understanding (Ventegodt et al., 2003).

Sometimes quality of life is characterized by order: biologically, psychologically, socially, and religiously (Mason, 2018; Ventegodt et al., 2003). On the other hand, it should be in accordance with certain spiritual and religion ideals, encompassing experiences free from psychological distress, for instance being afflicted by no depression or anxiety, being optimistic and living a highly significant life (Mason, 2018). Spiritual experiences can include aesthetic thrills, satisfaction with achievements, adjustment to one's self or to others, feelings of self-respective and human dignity that pertains to the unknown future, happiness, elation, gregariousness, empathy, sympathy, morals, mental health, optimism, and a feeling of identifying with nature (Moberg, 1971, p. 1).

Concluding, the QoL theory discussed above is adapted to be used as a theoretical foundation to underpin this study due to its concept being relevant for achieving the study aim. Especially, in relation to residents' quality of life, its three crucial components adapted for this study are: first, subjective quality of life, second, objective quality of life, and third, existential quality of life (Baker, 2001; Creswell, 2007).

Methodology

To achieve the research main aim an exploratory study on proper methodology was performed. A qualitative approach was employed, and in-depth interview technique was used for the data collecting process, while content analysis technique being used to analysis data. The participants from the CBT destination were in Loei Province, Northeast Thailand. The village named "Chiang Khan" was selected for this study because its uniqueness area. For instance, its province is home to some of the most majestic mountains in Thailand, such as Phu Kradueng (1,316m), Phu Luang (1,351m) and Phu Ruea (1,365m). Although Loei offers numerous outdoor activities and fascinating cultural attractions, tourism is relatively new to the province, and the area is very unspoilt and it became very fascinating tourist attraction, causing tourist numbers to gradually increase over the years.

The participants were selected due to the unique and multiple socio-cultural background of the destination. Especially because since the nineteenth century Lao people crossed the Mekong to flee from the French who were colonizing Laos, the Vietnamese and Chinese following soon after (Tourism Authority of Thailand, N.D). As a result, the CBT's rich history, cultural heritage, and uniqueness of the local's cultural background will have a wide range of tourism resources to provide sufficient data (Jeon et al., 2016). Moreover, residents will certainly hold authentic data and genuine experience in dealing with a variety of situations while living in the CBT destination (Jeon et al., 2016). They will hold rich data, particularly when dealing with different kinds of situations and a variety of visitor backgrounds; hence the rich data relevant to the quality of life will help to achieve the research aim. The data was collected between June-August 2021 and content analysis technique was used to analyse data. The interview questions were asked about socio-demographics, backgrounds, the CBT environment in general, the impact on economics, and the positive and negative impacts from the CBT development.

Participant Profiles

A total of 30 participants were interviewed: 9 males and 21 females. Most of them were born and are still living within this CBT village. The age group ranged between 78 and 31 years, and those around 40 years of age were the majority among the participants. All of them have their own businesses, including homestay business (11), street food vendor (7), farmer, including organic farm (8), and one each for retired, owning clothing store, and landlord respectively (as seen in Table 5.1).

Results

This chapter primarily investigated resident quality of life toward CBT development. The findings were based on interview questions, for example, "What is resident opinion on the community environment and atmosphere?" another question was, "What are the positive and negative impacts of CBT development?" Results are discussed below.

When asked what the residents thought about the environment and atmosphere within the community, the overall results revealed that the community environment and atmosphere had changed significantly since their community became a tourism destination. As an example a participant said that previously the village was so quiet, with hardly any loud noises around the area. Their authentic way of living was very independent (working on farms, respecting each other's privacy, and helping with unity the community when organizing a festival and traditional local culture). The residents usually went to bed very early because of living in the countryside, and there was nothing much to do after dark [ID6, ID11]. Respondents said the locals are very friendly, kind, and helpful to each other [ID14]. The community's surrounding areas obviously now has a lot of garbage within the village and along the roads. Increased traffic jams have worsened pollution especially with local roads that are usually small and narrow, when more cars and tourists floc to the place it becomes crowded; life is busier than before [ID 3, 4, 5, 6, 8, 9, 10, 11, 12, 13]. For example, one participant expressed:

> *"We are faced with a waste problem, it has increased significantly since the development of the tourism in our village, I always see garbage overflowing along the tourist walkway; the traffic never jammed before, but now I often see it is quite bad, polluted and crowded." [ID 5] Another said, "Since our village became a tourist place it is so noisy, especially at lunch time, lots of noise from motorcycles that is sometimes very disturbing, especially at*

Table 5.1 Participant profiles from Chiang Khan, Loei province

ID No.	*Age*	*Gender*	*Background living in this village*	*Own business type*
ID1	55	Female	Since they were born	Homestay
ID2	60	Male	Since they were born	Homestay
ID3	56	Female	Since they were born	Rubber farm
ID4	49	Female	Since they were born	Street food vender
ID5	39	Female	Since they were born	Local clothing store
ID6	41	Female	Since they were born	Street food vender
ID7	47	Female	Since they were born	Homestay / Restaurant
ID8	66	Female	Since they were born	Homestay
ID9	71	Female	Since they were born	Homestay
ID10	78	Female	Since they were born	Landlord
ID11	42	Female	Since they were born	Resort
ID12	39	Female	Since they were born	Street food vender
ID13	63	Male	Since they were born	Street food vender
ID14	48	Female	Since they were born	Street food vender
ID15	62	Male	Since they were born	Retirement
ID16	54	Female	Since they were born	Organic farmer
ID17	69	Female	Since they were born	Organic farmer
ID18	47	Female	Since they were born	Organic farmer
ID19	77	Female	Since they were born	Farmer
ID20	54	Male	Since they were born	Farmer
ID21	41	Male	Since they were born	Fisherman/farmer
ID22	40	Male	Since they were born	Organic farmer
ID23	35	Male	Since they were born	Homestay/Restaurant
ID24	31	Male	Since they were born	Homestay /Restaurant
ID25	59	Female	Since they were born	Restaurant
ID26	46	Male	Since they were born	Resort
ID27	48	Female	Since they were born	Street food vender
ID28	49	Female	Since they were born	Street food vender
ID29	73	Female	Since they were born	Homestay
ID30	71	Female	Since they were born	Homestay

> *night when they are running around the place; sometimes this loud noise also disturbs our guests late at night." [ID 7]*

Another question asked, "What are the positive and negative impacts on resident quality of life after CBT has been developed?" The participants revealed that the CBT development has both positive and negative impacts on their quality of life. The positive impact can include, 1) Economic impact, for instance, household income, joblessness decreased, family status has improved. 2) Family status improved, such as residents return home to start a business and be with family members. 3) Government support; the local government has paid more attention and supported the community with the necessary tourism resources, such as tourism facilities, accessibility, and tourism sectors involvement in helping community tourism development. 4) Environmental improvement, for example, the authority is stricter with environmental safety and cleanliness, more regulations are put in place, waste management strategy is implemented, such as the co-operation from the community on collecting garbage, and tourist ethics and responsibility help to keep the place nice and clean. 5) Health well-being: health care has become one crucial job to be concerned about as the government assign more staff from the health department to check up and focus on community health well-being. Lastly, 6) Socio-cultural values are preserved and sustainable, which in turn makes locals proud of their own local culture and traditions.

For the positive impacts on the economy including household income, they said that the financial situation was much better since the village became a tourism destination; they earned more money from doing some businesses, and CBT created more jobs within the community. The residents' incomes increased significantly, ranging from 60%-100% compared to their previous incomes, as they were mainly farmers, or worked in the rice fields, or made money from being a fisherman.

> *"I was originally born in this village, and I had to leave my hometown because not much work was available, and my earnings were not enough for living. I used to be a farmer, planting some rice and seasonal vegetables, but now I open my house selling seafood to tourists." [ID 4] Another said, "I was born here, and worked in Bangkok for a long time, but when my village become CBT, I returned to my village and started my business., I turned my house into a homestay and restaurant selling some local food, and local traditional clothes" [ID7].*

For the positive impact on the family status, the participants revealed that the CBT development can make a resident return home because there is now an opportunity to earn better money from their own hometown by starting a business. Previously, some residents were forced to leave their own home for a better future, a good life and better work; forcing these people to live far away from home and family. Therefore, CBT development has helped residents to be reunited with family, bringing a good quality of life to the family and in turn has lead to a healthy society. For instance, a participant revealed:

> *The residents who worked in the other provinces begin to return home and previously they considered selling their properties, but now I see them slowly turn their houses into some business; I think we all see such a good*

> *opportunity to make money from our own backyard." [ID6] Another said, "The good thing of CBT development is I earned more money, and I saw that many residents who worked in the other regions returned home because they saw the way to start some new businesses and a good chance to earn more money here. [ID11].*

For the positive impact on resident health well-being, it was found that since the community became CBT, especially during the COVID-19 pandemic, the residents received more attention from the government which provided healthcare staff to give some special advice and look after their health by regularly visiting people in the village, as some participants expressed:

> *The health care service system was improved dramatically. The government strongly supported our health well-being and regularly we see many health care staff come to our village conducting medical checkups, especially during the COVID-19 pandemic. They come to look after almost everyone in the village and provide vaccines for us all. [ID8]*

Another participant said,

> *The economics were good, better, and there was more money and satisfaction. Especially our health has been well looked after by the government, and I feel safe and secure when the government is seriously looking after our health consciously. [ID 9]*

Furthermore, a participant revealed that the CBT development helped to promote local culture and traditional lifestyle; the development makes local villagers proud of their authentic local culture. For one example a participant said, "The government launched a campaign to promote and preserve the local lifestyle, festivals, and authentic traditional housing style." [ID6, 11]

At the same time, the development of CBT has also created some negative impacts for the community. In terms of economics, the cost of living is high, and with more businesses in the community sometimes there are disputes between local owners; the situation makes people become more selfish, the friendly talk like in the old days has vanished. They said we earned more money, but we also lost some friendship and privacy. There will always be new visitors coming and going in their village, and they must be aware of who is good or bad, but they can't stop them coming, as tourists bring money into the community.

Residents who do not work within the tourism and hospitality services such as the locals who work for government, people who are retired, and some who work in private companies feel that the development of CBT has had no impact on their quality of life. Conversely, some said that the CBT has had a negative impact for them, especially on the living costs, as everything has increased, obviously because wherever tourism is developed living costs increase.

Another negative impact toward resident quality of life is related to community and resident health well-being. For example, participants said that their lifestyle has changed, they have lost their work – life balance and sleeping time has become unpredictable due to tourist and customer activity, and they must look after them well as it is part of the business. If good service is provided then there will intention to return, and business will be on-going. [ID6,7,8,11] For instance a participant asserted:

> *We have many tourists visiting our village and it is good for us when tourists spent more money, it is good for our business. I have homestay and yeah! I earn more money than before. On the contrary, I often don't have proper sleeping time as sometimes I go to bed very late, as I must look after my guest. So, the time to go to bed becomes unpredictable, no early nights like before the CBT development when I could set my sleeping time. [ID1]*

Another negative impact was the change in the local lifestyle. The peaceful and quiet community has changed its local life and become a noisy place. Typically, for the local lifestyle in the country in the evening after most of villagers return home from physical work on the farm or in the rice field, they will get tired, and maybe after dinner go to bed early. However, CBT development has changed the way of people's life; the little village becomes crowded with nightlife activity, a colorful and noisy place making the community lively, and waking up some people who want to sleep [ID 6, 7,11].

In summary, the overall results revealed both positive and negative impacts for resident quality of life toward CBT development. However, the analysis of the results, of the either positive or negative impacts can be used as critical points that have created guidelines to develop CBT and develop or improve resident quality of life significantly.

Conclusion and Discussion

The development of CBT can increase both resident quality of life and local business development within local communities. The study yielded findings on the idea that the development of CBT must incorporate goal setting, planning, implementing and evaluating, thus helping to overcome any negative impact from the development (Mathew & Sreejesh, 2017; Okazaki, 2008; Pinel, 1999). In respond to its main aim, this chapter seeks to know resident quality of life and community based tourism destination development and to achieve the aim, three research objectives were stated. First, to study resident subjective quality of life toward CBT development; second, to study resident objective quality of life toward CBT development; and third, to study resident existential quality of life toward CBT development. The analysis theoretically confirmed (Ventegodt et al., 2003), that the resident quality of life and community based tourism development model approaches comprises three components: first, resident subjective quality of life; second, resident objective quality of life; third, resident existential quality of life.

First, the findings identified the subjective quality of life referring to life satisfaction and overall happiness (Sirgy, 2010). An effective way to this approach is firstly to create a CBT atmosphere, especially an authentic local lifestyle such as peacefulness, friendliness, cleanliness, and avoid making the place becoming an overcrowded, over-tourism business destination. This approach can be accomplished by all the relevant stakeholders involved in community ethics; it is the responsibility of all businesses, locals, and tourists. It is everyone's duty to properly manage a community, particularly in keeping the place nice and clean, living in harmony with safety and security, ensuring road safety, traffic problems and pollution strategies are put in place. In addition, an effective way to create good ethics is to build awareness and consciousness by consistent communication among stakeholders regarding the community vision and values; the message must pass on from top to bottom, bottom to top. This includes stakeholders from non-profit and profit organizations, the community at all levels, local business owners and tourists.

In conclusion, the subjective quality of life must focus on "ethical duties," and doing the right thing will make everyone happy and satisfied with life. Consequently, it helps foster a sense of well-being, as all will feel part of the community while creating a good social life. Acts of kindness and friendliness are very important for residents, keeping everyone involved satisfied and happy with the CBT development.

Second, the findings also identified suitable concepts for the objective quality of life; it is about external factors that will impact resident quality of life. It refers to resident life fulfillment of needs or wants, how they are perceived by outside world and influenced by the community cultural norms. Thus, this approach can be accomplished by residents' perception of economic, social and environmental conditions (Jeon et al., 2016). The current results highlighted those factors to be effective for resident quality of life due to the CBT development: first, the economic development process, simultaneously improving the household income status, and financial improvement. Second, family status improvement, this phenomenon can be effective when people can work and stay near family members or establish a tourism and hospitality business at their own home. Third, government support for CBT development impacts resident quality of life. Tourism development will not be successful without the support of local authority and all concerned government departments. Especially, the health department must give priority to health well-being support for residents and community concerns. Fourth, environmental improvement: an environment sustainability strategy must be implemented when developing CBT to create resident quality of life; crucially, it is the stakeholders' responsibility. Fifth is to implement a health well-being strategy, especially with the main support coming directly from the government and health department. Sixth, socio-cultural awareness: it can be effective when it's creating a sense of identity, generating a feeling of belonging and pride in their own local culture, for example when the village chief, local authority, or relevant management show concern about cultural conservation for keeping the place authentic. This can be accomplished by conducting cultural street walks, dressing up in local traditional clothes, preserving the housing style by building in their own traditional style only, and not building modern houses in reserved areas.

Third, the existential quality of life: the study findings show further evidence of the role of the existential aspect within a quality-of-life development. The existential quality of life relates to what people believe about life and reality; it involves spirituality and the assumption that all humans are seeking for a meaning in life (Davison & Jhangri, 2010). Some spiritual experiences include aesthetic thrills, satisfaction with achievements, adjustment to oneself or others, or a feeling of self-respect and human dignity (Moberg, 1971). The role of spiritual experience can function as an existential quality of life catalyst when guided by stakeholders' collaboration and community strategy plans, particularly in relation to improved local culture traditional related to spiritual believed, and value. For instance, the festival and traditions related to spiritual well-being need to be well maintained, and the reservation of a worship place of village is necessary. This links in turn to resident happiness, satisfaction, mental well-being, and meaningful life, ultimately reaching their self-actualization goal. Moreover, an effective way for an existential quality of life is for residents to live in a harmonious society, free from psychological distress, especially no depression or anxiety when their internal life has been fulfilled with peace. Living in a natural environment, respecting individual lives, and own spiritual beliefs and good ethics can foster an optimistic and highly meaningful life, particularly practising the spiritual retreat activity from local spiritual culture, such as making merit and celebrating on a Buddhism day or other spiritual retreat not related to religion (Mason, 2018).

Management Implication

The relevant stakeholders in the community all play an important role in successfully creating resident quality of life toward CBT development. The theoretical draw out from this study points to a possible strategy for related stakeholders to apply and develop to suit their own businesses. It is also to be used as a guideline for tourism planning and strategy in all destinations if there is a need for development. Most importantly, the strategy will be successful with government support, such as for the development of tourism infrastructure, including environmental sustainability, building attractive communities, tourism facilities of high standard, and good quality of accommodation or homestay. Access to the community needs to be sustainable, such as building more roads with a variety of transportation modes, not overpriced, affordable for all tourist groups. The development of various tourism activities must be authentic, not causing any negative impact on the community, as such the activity must be suitable with the local environment, for example a stakeholder may bring scooters for tourists to rent and ride within the locality of the village and it may be good for the tourists and business owners, but it also can cause accidents, is noisy and it doesn't fit with the local village atmosphere. The amenities must be sufficient, and serve both locals and tourists, especially with the establishment of a facility or centre for health and well-being in the village. Moreover, an overall strategy for the quality of life and CBT development must consider sustainability and responsibility, as having good ethics and morals are most crucial for our society and the global environmental world. Finally, without human ethics the place will not be a pleasant place to live. Thus, it is a duty of all citizens to make the village a better place to live, and in particular all must respect individual belief, the environment, and people's own spiritual beliefs rooter either in natural or supernatural faiths.

References

Ali, M. B., Quaddus, M., Rabbanee, F. K., & Shanka, T. (2022). Community participation and quality of life in nature-based tourism: Exploring the antecedents and moderators. *Journal of Hospitality & Tourism Research*, 46(3), 630–661.

Apollo, M., & Andreychouk, V. (2022). *Mountaineering Adventure Tourism and Local Communities: Social, Environmental and Economic Interactions.* Edward Elgar Publishing. https://doi.org/10.4337/9781802209389.

Apollo, M., & Moolio, P. (2021). *Poverty and Development: Problems and Prospects*: Channel View Publications. https://doi.org/10.21832/9781845418489.

Baker, M. J. (2001). Selecting a Research Methodology. *The Marketing Review* 1(3), 373–397.

Blunsdon, B., & Davern, M. (2007). Measuring wellness through interdisciplinary community development: Linking the physical, economic and social environment. *Journal of Community Practice*, 15(1–2),217–238. https://doi.org/10.1300/J125v15n01_10.

Carley, M. J. (1979). Social theory and models in social indicator research. *International Journal of Social Economics*, 6(1), 33–44.

Chan, K. (2018). Is religious and existential well-being important in quality of life in Hong Kong Chinese? *The Social Science Journal*, 55(3), 273–283. https://doi.org/10.1016/j.soscij.2018.02.012.

Chok, S., Macbeth, J., & Warren, C. (2007). Tourism as a tool for poverty alleviation: A critical analysis of 'pro-poor tourism' and implications for sustainability. *Current Issues in Tourism*, 10(2–3), 144–165. https://doi.org/10.2167/cit303.

Creswell, J. W. (2007). *An Introduction to Mixed Methods Research.* SSP, University of Nebraska-Lincoln.

Curcija, M., Breakey, N., & Driml, S. (2019). Development of a conflict management model as a tool for improved project outcomes in community based tourism. *Tourism Management*, 70, 341–354. https://doi.org/10.1016/j.tourman.2018.08.016.

Davison, S. N., & Jhangri, G. S. (2010). Existential and religious dimensions of spirituality and their relationship with health-related quality of life in chronic kidney disease. *Clinical journal of the American Society of Nephrology: CJASN*, 5(11), 1969.

Delita, F., Berutu, N., Tanjung, F., & Rambe, T. (2024). Resident's Attitudes Toward Tourism Development in Simanindo Subsdistrict Samosir Regency. *KnE Social Sciences*, 1165–1180. https://doi.org/10.18502/kss.v9i2.14932.

Donald, A. (1998). *What is quality of life?* (Vol. 1): Hayward Medical Communications.

Dunn, H. L. (1959). What high-level wellness means. *Canadian Journal of Public Health/Revue Canadienne de Sante'e Publique*, 50(11), 447–457.

Estes, R. J., & Sirgy, M. J. (2019). Global advances in quality of life and well-being: Past, present, and future. *Social Indicators Research*, 141, 1137–1164. https://doi.org/10.1007/s11205-018-1869-4.

FDHA. (2016). *Quality of Life in the Cities 2016*. Neuchatel, Switzerland: Federal Statistical Office Retrieved from https://www.bfs.admin.ch/bfs/en/home/statistics/cross-sectional-topics/city-statistics/indicators-quality-life.assetdetail.1401994.html.

Gohori, O., & van der Merwe, P. (2024). Barriers to community participation in Zimbabwe's community-based tourism projects. *Tourism Recreation Research*, 49 (1), 91–104. https://doi.org/10.1080/02508281.2021.1989654.

Goodwin, H., & Santilli, R. (2009). Community-based tourism: A success. *ICRT Occasional paper*, 11(1), 37.

Goodwin, H., & Santilli, R. (2009). *Community-based Tourism: A Success?*. University of Greenwich.

Gregoire, T. K. (2002). The validity of a social indicator approach to substance misuse needs assessment. *Substance Use & Misuse*, 37(3), 357–380. https://doi.org/10.1081/JA-120002483.

Harrison, D., & Schipani, S. (2007). Lao tourism and poverty alleviation: Community-based tourism and the private sector. *Current Issues in Tourism*, 10(2–3),194–230. https://doi.org/10.2167/cit310.0.

Javed, S., Javed, S., & Khan, A. (2016). Effect of education on quality of life and well being. *The International Journal of Indian Psychology*, 3(3), 119–128. doi:10.25215/0304.053.

Jeon, M. M., Kang, M., & Desmarais, E. (2016). Residents' perceived quality of life in a cultural-heritage tourism destination. *Applied Research in Quality of Life*, 11, 105–123. https://doi.org/10.1007/s11482-014-9357-8.

Khazaeipul, M., & Ramezanzadeh Lasboyee, M. (2023). The Community-based Tourism Model of Mountain Villages with the Grounded Theory Approach (Case Study: Kojur District, Nowshahr County). *Journal of Research and Rural Planning*, 12(1), 61–76. doi:10.22067/jrrp.v12i1.2212-1068.

Kim, K., Uysal, M., & Sirgy, M. J. (2013). How Does Tourism in a Community Impact the Quality of life of Community Residents?. *Tourism Management* 36, 527–540. https://doi.org/10.1016/j.tourman.2012.09.005.

Land, K. C., & Spilerman, S. (1975). Social indicator models: An Overview. In C. Kenneth and L. Seymour Spilerman (Eds.), *Social Indicator Model*(5-36). Russell Sage Foundation.

López-Guzmán, T., Sánchez-Cañizares, S., & Pavón, V. (2011). Community-based Tourism in Developing Countries: a Case Study. *Tourismos* 6(1), 69-84.

Makandwa, G., Makudza, F., & Muparangi, S. (2023). Women's Participation in Community-Based Tourism During the COVID-19 Pandemic: A Case of Sengwe Community in Zimbabwe. In K. Dube, O. L. Kupika and D. Chikodzi *(Eds.), COVID-19, Tourist Destinations and Prospects for Recovery: Volume Three: A South African and Zimbabwean Perspective* (115–131). Springer.

Marans, R. W., & Stimson, R. J. (2011). *Investigating quality of urban life: Theory, methods, and empirical research* (Vol. 45). Springer .

Mason, H. D. (2018). The relationship between existential attitudes and professional quality of life among nursing students. *Journal of Psychology in Africa*, 28(3), 233–236. https://doi.org/10.1080/14330237.2018.1435049.

Mathew, P. V. (2022). Sustainable tourism development: discerning the impact of responsible tourism on community well-being. *Journal of Hospitality and Tourism Insights*, 5(5), 987–1001. doi:10.1108/JHTI-02-2021-0052.

Mathew, P. V., & Sreejesh, S. (2017). Impact of responsible tourism on destination sustainability and quality of life of community in tourism destinations. *Journal of Hospitality and Tourism Management*, 31, 83–89. https://doi.org/10.1016/j.jhtm.2016.10.001.

Moberg, D. O. (1971). *Spiritual well-being: Background.* Paper presented at the Paper presented at the Aging National Planning Board.

Nadee, P., & Ashton, S. (2021, 13–15 September). *Residence Attitude and Quality of Life Post Community-Based Tourism Development in Thailand.* Paper presented at the Beyond Smart Tourism Inspired by People, Balanced by Nature and Driven by Technology Virtual Conference.

Nkemngu, A. (2015). Quality of life and tourism impacts: a community perspective. *African Journal of Hospitality, Tourism and Leisure*, 4(1), 1–13.

Norman, A., & Pokorny, J. J. (2017). Meditation retreats: Spiritual tourism well-being interventions. *Tourism Management Perspectives*, 24, 201–207. https://doi.org/10.1016/j.tmp.2017.07.012.

Nugroho, P., & Numata, S. (2021). Changes in residents' attitudes toward community-based tourism through destination development in Gunung Ciremai national park, Indonesia. *Tourism Recreation Research*, 46(3), 403–421. https://doi.org/10.1080/02508281.2020.1808753.

Nur, A. Y., Istiqomah, N. m., Priambodo, M. P., Puspasari, E. Y., Sidi, F., & Marzanah, A. J. (2021). *Community Based Rural Tourism (CBRT): The Impact on Local Residents Quality of Life in Indonesia.* Paper presented at the Paper presented at the E3S Web of Conferences.

Oei, T. P., & McAlinden, N. M. (2014). Changes in quality of life following group CBT for anxiety and depression in a psychiatric outpatient clinic. *Psychiatry Research*, 220(3), 1012–1018. https://doi.org/10.1016/j.psychres.2014.08.036.

Okazaki, E. (2008). A community-based Tourism Model: Its Conception and use. *Journal of Sustainable Tourism*, 16(5), 511–529. https://doi.org/10.1080/09669580802159594.

Pinel, D. P. (1999). Create a good fit: A community-based tourism planning model. In M. L. Miller, J. Auyong, & N. P. Hadley (Eds.), *Proceedings of the 1999 international symposium on coastal and marine tourism* (277–286). Vancouver. Retrieved from http://nsgl.gso.uri.edu/washu/washuw99003/28-Pinel.pdf.

Qu, C., Zhang, C., Shen, S., & Olsen, D. H. (2023). Heritage conservation and communities' sense of deprivation in tourism: the case of the Hani community in Yunnan, China. *Tourism Geographies*, 25(2–3),881–898. https://doi.org/10.1080/14616688.2021.2016936.

Raphael, D., Renwick, R., Brown, I., Steinmetz, B., Sehdev, H., & Phillips, S. (2001). Making the links between community structure and individual well-being: community quality of life in Riverdale, Toronto, Canada. *Health & Place*, 7(3), 179–196. doi:10.1016/s1353-8292(01)00008-9

Sebele, L. S. (2010). Community-based Tourism Ventures, Benefits and Challenges: Khama Rhino Sanctuary Trust, Central District, Botswana. *Tourism Management* 31(1), 136–146. https://doi.org/10.1016/j.tourman.2009.01.005.

Sirgy, M. J. (2010). Toward a quality-of-life theory of leisure travel satisfaction. *Journal of Travel Research*, 49(2), 246–260. https://doi.org/10.1177/0047287509337416.

Theofilou, P. (1995). *Quality of Life: Definition and Measurement Europe's Journal of Psychology*, 9(1), 150–162.

Tourism Authority of Thailand. (N.D). *Experience Chiang Khan with your heart* Retrieved from https://www.thailandtravel.or.jp/common/pdf/Experience_chiangkhan.pdf.

Uysal, M., Sirgy, M. J., Woo, E., & Kim, H. L. (2016). Quality of Life (QOL) and Well-being Research in Tourism. *Tourism Management*, 53, 244–261. https://doi.org/10.1016/j.tourman.2015.07.013.

Ventegodt, S., Merrick, J., & Andersen, N. J. (2003). Quality of life theory I. The IQOL theory: an integrative theory of the global quality of life concept. *The Scientific WorldJjournal*, 3, 1030–1040.

Wani, M. D., Batool, N., Dada, Z. A., & Shah, S. A. (2024). Investigating the impact of community-based tourism on the residents' quality of life and their support for tourism. *Community Development*, 55(1), 138–159. https://doi.org/10.1080/15575330.2023.2272271.

Yeap, P.-F., & Liow, M. L. S. (2024). Tourist walkability and sustainable community-based tourism: conceptual framework and strategic model. *International Journal of Tourism Cities*, 10(1), 78-104. https://doi.org/10.1108/IJTC-05-2022-0117.

Yunikawati, N. A., Istiqomah, N. m., Priambodo, M. P., Puspasari, E. Y., Sidi, F., & Marzanah, A. J. (2021). *Community Based Rural Tourism (CBRT): The Impact on Local Residents Quality of Life in Indonesia*. Paper presented at the Paper presented at the E3S Web of Conferences.

Yusof, Y., Ibrahim, Y., Muda, M. S., & Amin, W. A. A. W. M. (2012). Community based tourism and quality of life. *Review of Integrative Business and Economics Research*, 1(1), 336-346.

6 Mountain Dark Tourism

A Development Pathway in Post-conflict Zones in Colombia

James Pérez-Morón, Ulf Thoene and Roberto García Alonso

Introduction

The allure of mountains as exotic destinations and the tourism industry has a dynamic interaction, resulting in a substantial economic impact. According to the Food and Agriculture Organization of the United Nations and the World Tourism Organization (2023), international arrivals to mountainous regions generate an estimated equivalent of $195–375 million USD. These natural highland landscapes are among the most sought-after tourist choices (Nepal & Chipeniuk, 2006). However, it is crucial to recognize that these mountains are also home to some of the world's poorest populations (also known as the base of the pyramid) (Prahalad & Hart, 2010), and 25 per cent of the world´s terrestrial biodiversity (Apollo & Wengel, 2022).

Activities such as hiking, trekking, and climbing (Apollo, 2017) are popular tourist attractions, offering opportunities for adventure and exploration in unique settings. Additionally, mountains often hold religious and cultural significance, attracting visitors seeking spiritual experiences or cultural immersion. Despite the appeal and multifaceted interaction of mountain tourism, there are barriers to achieving sustainable practices in these regions (Musa et al., 2004; Stefano & Letey, 2019), including the accessibility of the destinations and the real accessibility to the mountains themselves (Apollo, 2017).

Mountains serve tourist destinations and have a darker side to their history. Historically, mountains have been scenes of armed conflicts worldwide, often acting as safe havens for criminal elements (Ospina, 2023) or as scenarios that suffered the environmental consequences of the war, devastating, particularly for peasant farmers, indigenous groups, and marginalized communities (Olson & Speidel, 2023; Sayed et al., 2023). While the literature extensively addresses mountaineering accessibility (Apollo, 2017; Apollo & Rettinger, 2019; Garavaglia & Diolaiuti, 2012; Palomo, 2017; Witte et al., 2023), there is a significant research gap when it comes to social accessibility to mountainous regions affected by armed conflicts and the political aspects involved.

Simultaneously, the literature falls short in explaining the phenomenon of dark tourism as a potential approach for the economic, social, and environmental benefit of mountain dwellers. Dark tourism refers to tourists' interest in visiting places associated with death, macabre events, and humans (Bowman &

DOI: 10.4324/9781003392583-7

Pezzullo, 2009; Stone & Sharpley, 2008). Consequently, the primary objective of this study is *to deepen the understanding of dark tourism in mountain regions as a suitable means of reintegrating former combatants of the FARC (Revolutionary Armed Forces of Colombia) and the victims of the Colombian armed conflict.*

To address this research gap, we adopt a mini-case study approach, following Stone & Grebenar′s mini-case study (2022) from the 2017 terrorist attack at the Ariana Grande concert and its subsequent "tragic placemaking" of Manchester, UK, focusing on a tourism cooperative formed by women who are victims of the FARC. This cooperative operates in Chalán, a mountainous region in Montes de María, located in the northern Caribbean region of Colombia, an area with a historical legacy of armed conflict, poverty, and illicit crops such as coca, marijuana, and poppy, with multiple guerrilla groups and other illegal armed groups coexisting.

Literature Review

This chapter draws on the dark and mountain tourism and armed conflict literature review analysis (Saini et al., 2022) following four stages. Out of the multiple journal ranking lists, including the Australian Business Deans Council (ABDC), the social sciences citation index (SSCI) of Clarivate Analytics (Web of Science-WOS), and Elsevier's Scopus, we opted to choose Scopus due to its wider coverage (Saini et al., 2022). To ensure the robustness of the Scopus results, we validated them with WOS data. We thoroughly analyzed the keywords "dark tourism" and "armed conflict" to construct the search formula. We included additional keywords such as "mountain tourism," "grief tourism," "thanatourism," "black tourism," "morbid tourism," "battlefield tourism," "war tourism," "warfare tourism," "danger zone tourism," and "conflict-heritage tourism." The search field chosen was "title, abstract and keywords," and Boolean operators "AND/OR" were utilized. This process resulted in the final search terms: "TITLE-ABS-KEY ('mountain tourism' AND 'dark tourism' AND 'armed conflict' AND 'grief tourism' AND 'thanatourism' AND 'black tourism' AND 'morbid tourism' AND 'battlefield tourism' AND 'war tourism' AND 'warfare tourism' AND 'danger zone tourism' AND 'conflict-heritage tourism')." The last search for this chapter was conducted on June 14, 2023.

We retrieved 755 documents (Figure 6.1), based on the keyword search conducted from 2000 to 2023. To refine the selection, we established inclusion and exclusion criteria. We focused solely on documents within the *"Business, Management, and Accounting"* subject category, resulting in 501 documents. We included articles and review articles while excluding other document types such as books, book chapters, conference papers, and letters. Furthermore, we only selected articles written in English, leading to a final set of 322 documents. We carefully examined the abstracts from this pool to identify those relevant to our criteria. Ultimately, we selected a subset of 21 articles. To facilitate analysis, all these articles were exported from Scopus in a comma-separated (.csv) file and further analyzed using Biblioshiny, a tool based on the R language.

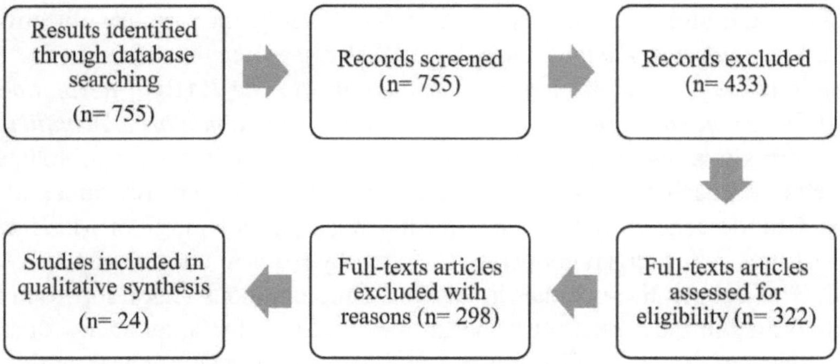

Figure 6.1 Flow chart of the study selection process

We selected Biblioshiny (Saini et al., 2022) as our chosen tool for literature analysis due to its diverse range of features. Utilizing Biblioshiny, we analyzed to determine the annual scientific production and identified the most influential sources in the field. Additionally, we utilized a three-field plot to visually represent the interplay between keywords, authors, and journals, providing a comprehensive overview of the research landscape. This approach enables us to gain deeper insights into the trends and patterns within literature and facilitates a more comprehensive understanding of the topic at hand.

The annual production of 24 documents from 2010 to 2023. with a yearly growth rate of 11.25 and an 18.1 average citation per article. These 24 articles were published in 14 journals, written by 64 authors and 1,386 references.

An article-wise analysis reveals the field's top 10 most cited documents (Table 6.1). One notable study by Lemelin et al. (2010) garnered 169 citations, introducing the concept of "doom tourism" or "last-chance tourism." This groundbreaking work explored why tourists are attracted to environmentally damaged or endangered destinations. Another highly cited article by Kumar et al. (2020) received 37 citations and provided a comprehensive review of key themes in tourism, including dark tourism, in commemoration of the 15th anniversary of the Journal of Heritage Tourism. We emphasized the growing attention that dark tourism was receiving in contemporary tourism research.

Furthermore, several articles focused on specific aspects of dark tourism, such as massacre tourism (Frew, 2012), tourist motivations to visit post-disaster sites (Chen et al., 2017; Qian et al., 2017; Wang et al., 2021; Wu & Cheng, 2018) grim and haunted tourism (Cornell et al., 2019), and tourism involving death rituals and cremation ashes (Light et al., 2023). However, none of these articles explicitly addressed the potential for dark tourism to benefit mountain dwellers or explored its direct implications for local communities.

Only three articles qualitatively examined the connection between dark tourism and armed conflict. Mansfeld and Korman (2015) investigated tourism in conflict-affected Israeli border areas, while Kamber et al. (2016) explored tourist motivations for war-related sites in Sarajevo. Another study

Table 6.1 Top-ten most cited documents

No.	*Author/year*	*Title*	*Total Citations*
1	Lemelin et al. (2010)	Last-chance tourism: The boom, doom, and gloom of visiting vanishing destinations	167
2	Kumar et al. (2020)	The Journal of Heritage Tourism: a bibliometric overview since its inception	37
3	Frew (2012)	Interpretation of a sensitive heritage site: The Port Arthur Memorial Garden, Tasmania	34
4	Prayag (2018)	Domestic tourists to Elmina Castle, Ghana: motivation, tourism impacts, place attachment, and satisfaction	31
5	Wu & Cheng (2018)	What drives supportive intentions toward a dark tourism site?	29
6	Qian et al. (2017)	Hit close to home: the moderating effects of past experiences on tourists' on-site experiences and behavioral intention in post-earthquake site	19
7	Chen et al. (2017)	Influence of Place Identity on Residents' Attitudes to Dark Tourism	13
8	Qian et al. (2022)	Dark tourism destinations: the relationships between tourists' on-site experience, destination image, and behavioral intention	7
9	Wang et al. (2021)	Interpreting disaster: How interpretation types predict tourist satisfaction and Loyalty to dark tourism sites	7
10	Skinner (2018)	Plymouth, Montserrat: apocalyptic dark tourism at the Pompeii of the Caribbean	7

by Applboim and Poria (2020) focused on Holocaust tourism in heritage sites in Poland. Although these articles shed light on the relationship between dark tourism and armed conflict, they did not explicitly discuss the potential for utilizing dark tourism to benefit mountain communities.

These findings underscore the existing research gaps and highlight the need for further investigation into the potential benefits of dark tourism for mountain dwellers. By examining the experiences of former combatants and victims of armed conflict, we can explore how dark tourism might contribute to these communities' economic, social, and environmental well-being.

The three-field plot, depicted in Figure 6.2, utilizes Sankey diagrams to visualize the relationship between countries, keywords, and journals (Saini et al., 2022). In this plot, the box size indicates the frequency of occurrences. Analysis of the plot reveals interesting patterns. Asian authors, primarily from China, Thailand, and the Philippines, commonly employ keywords such as "dark tourism" and "tourism." On the other hand, American authors from the USA and Canada do not only focus on "dark tourism" but also explore

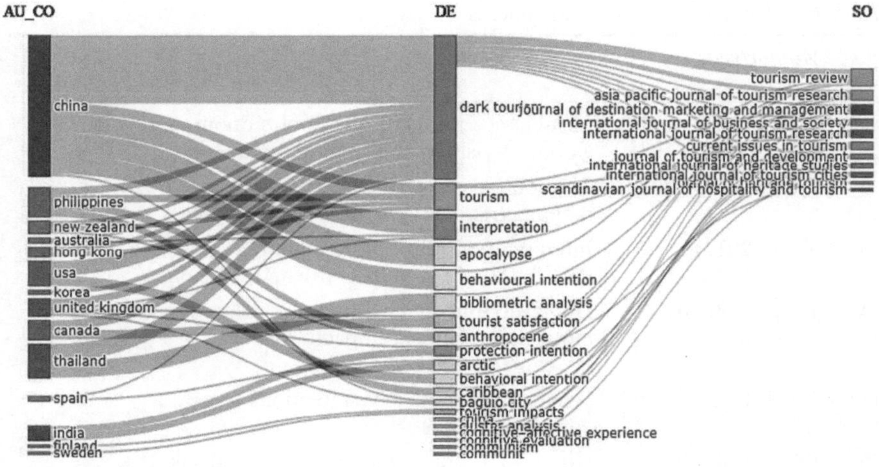

Figure 6.2 Three-field plot
Source: Biblioshiny

topics related to "tourist destinations" and "behavioral intentions." European authors, including those from Finland, Sweden, and the UK, tend to prioritize studying tourism impacts.

Examining the plot from a journal perspective, it becomes evident that the keyword "dark tourism" is prominently featured in publications such as Tourism Review, Asia Pacific Journal of Tourism Research, and Journal of Destination Marketing and Management. These journals demonstrate a specific interest in research related to dark tourism, emphasizing its significance in the field. The three-field plot highlights the varying research emphases across different regions and journals. It underscores the global interest in studying dark tourism while revealing distinct regional preferences and the diverse keywords explored in scholarly work.

Study Area

The area under consideration corresponds to *Chalán* which is characterized by its moderate altitude, typically ranging between 200 to 1,000 meters above sea level. This region is known for its rolling hills and lush landscapes, nestled within a relatively low mountain range that separates the basins of the Magdalena and San Jorge rivers. Since 2017, it has been designated as one of the Development Programs with Territorial Focus (PDET), which serves as a planning and management instrument for implementing the peace agreements signed between the Colombian State and the FARC in 2016. Chalán is part of a larger region known as Montes de María, encompassing municipalities from two departments: Bolívar and Sucre. In Bolívar, the municipalities include Marialabaja, San Juan Nepomuceno, San Jacinto, Zambrano, El Guamo, and El Carmen de

Bolívar. In Sucre, the municipalities are Morroa, Colosó, Tolú Viejo, Los Palmitos, Chalán, Corozal, and Ovejas. Montes de María is situated in the northern Caribbean region of Colombia, an area deeply impacted by the historical internal armed conflict, widespread poverty, and the cultivation of illicit crops such as coca, marijuana, and poppy. Various guerrilla groups and illegal armed factions have marked it.

Montes de María's subregion has endured the armed conflict's consequences for several decades, compounded by institutional weaknesses and the precariousness of land use planning. These factors have created fertile ground for high poverty rates and limited social and economic development, resulting in significant disparities in competitiveness, particularly in terms of infrastructure. The rural areas have borne the brunt of these impacts, exacerbated territorial inequalities, and widened gaps in social, economic, and gender development. Consequently, poverty rates are disproportionately higher in these areas, and the prospects for overcoming them are considerably more challenging, particularly for rural women. Hence, it is imperative to promote mechanisms that enable organized civil society to participate in decision-making processes and contribute to the reconstruction of the social fabric.

Chalán has a surface area of 40 km^2 and includes one municipal capital and four rural municipalities: *La Ceiba, Nuevo Manzanares, Desbarrancado y Montebello.* In 2022, it had 4,425 inhabitants (2,362 men and 2,063 women). 3.6% of its inhabitants are indigenous, 15.2% Afro-Colombians and 81.2% belong to non-ethnic minorities (Government of Sucre, 2022). *Chalán* has several unsatisfied needs, such as a decent house, unemployment, overcrowding, and public services, and their main economic activities are agriculture, small-scale livestock farming, and handicraft artisans.

Chalán is also a forest reserve with 6,452 hectares, mainly tropical. Its mountains are ideal for nature, eco-tourism, and birdwatching tourism, and its rich cultural heritage has allowed its cultural tourism (including dance, gastronomy, and handicrafts). Below, one can find the most important natural attractions:

- *Piececito de San Antonio:* Mountain Creek, also the center of religious acts
- *Alberca de Tonaya:* Natural water Sources in the mountain, ideally for hiking
- *Las Colinas:* Beautiful Waterfall that flows into a natural pool
- *El Sereno:* A 30-meter-high natural waterfall
- *Las Tinas:* Two natural pools
- *Arroyo Tonaya:* Mountain formations ideal for hiking, and fauna sightings
- *Arroyo Pitalcito:* Es un arroyo caudaloso y de aguas cristalinas

- *Cueva de los Murciélagos:* Rock formation, ideal for hiking
- *Caverna del Diablo:* Rock formation with a religious background

People from *Chalán* (*chalaneros*) are committed to reconstructing the social fabric and working in its community to prevent violence from returning to its territory. Community and sustainable tourism have emerged in this territory, and currently, *Chalán* has the following associations: (1) Asociación Ecológica y Turística de Chalán – ASOECOTURCHAL. Focus: Sustainable Tourism, restoration of the tropical dry forest and tourist trails; (2) ASOJUVENTUD. Focus: Restoration of forest; (3) Cooperativa Chalanera de Gestores de Paz de los Montes de María (formed by 16 ex-combatants from FARC). Focus: tourism, intensive cattle, improved yam and chili pepper growing; and (4) Excursiones Montes de María S.A.S. SOMAC. Focus: Tour Operators.

A case study focusing on Chalán, Montes de Maria region is crucial, as it presents a unique blend of historical, socio-political, and environmental elements. The control of this area by guerrilla groups has historically influenced its accessibility and development. Now in the post-conflict setting, the region experiences changes in governance and control, and mountaineering tourism emerges as a promising avenue for sustainable development and community empowerment. This case study could provide valuable insights into how regions with similar histories can leverage natural resources for economic and social progress while ensuring environmental conservation.

Cooperative to Build Peace in Chalán (COOBUPECHAN)

This tourism cooperative was established in Chalán in 2021, a territory plagued by guerrilla activity. The cooperative has directly benefited 38 ex-combatants, comprising 31 men and seven women. All participant and company names in the mini-case study are made anonymous by utilizing pseudonyms in order to protect sensitive information. One hundred and two families have experienced indirect benefits, and five touristic operators are also part of the cooperative. Initially, the cooperative was established as a livestock project, but it subsequently evolved into an ecotourism initiative focused on the high mountains. These mountains have remained untouched, as they served as a refuge for the guerrilla for an extended period, restricting access to outsiders. Consequently, foreigners and non-FARC civilians were deterred from visiting this area due to the risk of kidnapping, which was one of the FARC's primary revenue sources.

Historically, the guerrilla have demonstrated a tendency to protect the environment and refrain from deforestation, not out of love for the environment but due to strategic considerations. COOBUPECHAN, the tourism cooperative, has also aligned its practices with this approach. However, it is evident that, on a national scale, the peace accord has had mixed effects on the environment. Deforestation, cattle ranching, forest clearing, mining, and coca cultivation have increased in various regions (Cantillo & Garza, 2022 & Corredor-Garcia & López, 2023).

The ex-combatants of the FARC chose to remain in the territory as they were intimately familiar with the area and had enormous challenges to be part of society once again. This choice was influenced by the realization that residents and foreigners were strongly interested in visiting these unexplored and picturesque territories, which lacked basic amenities such as potable water, roads, infrastructure, bathrooms, and kitchens. Visitors sought to experience the lifestyle of combatants temporarily in the post-conflict era, and it was in response to this demand the concept of guerrilla jungle camps as mountain tourist destinations emerged. These camps and sleep huts replicated the living conditions of combatants during the internal conflict.

These guerrilla camps were primarily organized and led by ex-combatants along with their comrades, and they offered interested visitors, both local and foreign, the opportunity to immerse themselves in the *mise en scène* of the Colombian internal armed conflict (Bowman & Pezzullo, 2009). Visitors could assume the roles of ex-combatants or kidnapped civilians, becoming firsthand performers in recreating key elements from the conflict. The camps also served as a reconciliation tool, allowing former perpetrators and victims to work together. This collaborative environment facilitated ex-combatants' reintegration into civilian life and contributed to peace-building efforts.

By providing a space where visitors could directly engage with the experiences of combatants, the guerrilla jungle camps offered a unique form of tourism that combined elements of adventure tourism and historical reenactment. These camps did not only satisfy the curiosity of those interested in the Colombian conflict but also played a significant role in transforming ex-combatants from stigmatized figures to active participants in the reconciliation process. The interaction and cooperation between former combatants and visitors in these camps fostered understanding, empathy, and dialogue, promoting healing and the construction of lasting peace.

However, one drawback of this new endeavour was the lack of preparedness to receive visitors in an environmentally conscious manner. Initially, people were allowed to enter without consideration for the damage they might cause to the environment. Access to land has been a determining factor in rural poverty and conflicts.

Institutional weaknesses have hindered the formalization of tourism in these territories, which has favoured the ex-combatants but has not been as beneficial for the tourist sector. Even foreign direct investment in these types of projects has been slow to materialize, with the growth of such investments relying on national and international cooperation. As the informal authority in the territory, COOBUPECHAN ensures that mountain dwellers, refugees, and other conflict victims respect the environment and collaborate with them on these tourism initiatives rather than engaging in activities that may provide more immediate benefits but have long-term negative consequences for their territories.

On the other hand, the cooperative has expressed interest in continuing to work with the mountain dwellers, as they play a crucial role in providing a unique experience to visitors eager to immerse themselves in the guerrillero lifestyle. These mountain dwellers have allowed visitors to stay in their homes, provided meals, and acted as tour guides in the mountains and natural water bodies. For the mountain dwellers, this partnership presents an opportunity to establish connections with former combatants and generate income. They have had no alternative since ex-combatants wield authority in these territories, even in the post-conflict period, necessitating either alignment with them or facing potential dangers.

Armed or ex-armed actors set the rules and promoted migration to these territories, which led to the idea of recreating a guerrilla camp, simulating the experience of their former lives, not only for visitors but also to address feelings of homesickness. Without conducting market research or fully understanding the foundations of dark tourism, COOBUPECHAN proceeded with this idea based solely on the high demand expressed by tourists seeking to experience the guerrillero lifestyle. In doing so, they recognized the great opportunity it presented for visitors, mountain dwellers, and themselves – a true win-win situation.

Discussion and Conclusion

The literature review also uncovered various facets of dark tourism, such as massacre, post-disaster, grim, and haunted tourism, as well as the tourism of death rituals. Studies like those by Frew (2012) and Chen et al. (2017) contribute to understanding tourist motivations and the interpretation of sensitive heritage sites. However, a notable gap was identified: none of these studies explicitly addressed the potential of dark tourism in benefiting mountain communities or its direct implications for local populations. Only a few articles, like those by Mansfeld and Korman (2015) and Kamber et al. (2016), qualitatively examined the link between dark tourism and armed conflict, but they did not focus on the potential benefits for mountain communities. This gap underscores the need for more focused research in this area, particularly exploring how dark tourism could contribute to the economic, social, and environmental well-being of communities in mountain regions affected by conflict. The allure of mountains as exotic tourist destinations and the tourism industry's dynamic interaction has a significant economic impact, generating millions of dollars in revenue. However, it is essential to recognize that these mountainous regions are home to some of the world's poorest populations, highlighting the need to balance economic interests with the well-being of local mountain dwellers.

Mountain tourism offers a range of activities such as hiking, trekking, and climbing, attracting adventure-seeking tourists and those interested in cultural and spiritual experiences. However, achieving sustainable practices in these regions remains a challenge due to barriers like limited accessibility to the destinations and the mountains themselves. Efforts must ensure that mountain tourism activities are conducted responsibly to preserve the environment and benefit local communities (Palomo, 2017).

Mountains have served as tourist destinations and witnessed armed conflicts, resulting in devastating ecological consequences for marginalized communities. The existing literature addresses the accessibility of mountain regions but falls short in exploring the social accessibility of conflict-affected mountainous areas. Further research is needed to understand the political aspects involved and explore how tourism, including dark tourism, can contribute to the well-being of mountain dwellers affected by armed conflicts.

The study area of Chalán in Montes de María, Colombia, presents a unique case for examining dark tourism as a means of reintegrating former combatants and victims of the armed conflict. This region, marked by historical armed conflict, poverty, and illicit crop cultivation, provides an opportunity to explore the potential benefits of dark tourism for mountain communities. By studying the experiences of these communities, we can better understand how dark tourism can contribute to their economic, social, and environmental well-being. Despite Colombia not having high-mountain clubs affiliated with the International Mountaineering and Climbing Federation (UIAA), the country has started to foster the growth of mountaineering tourism.

Firstly, it is crucial to prioritize environmental conservation and adopt sustainable tourism practices. Mountain regions, often untouched and pristine, hold significant ecological value. Tourism initiatives should be designed to minimize negative impacts on the environment, including educating visitors about responsible tourism, implementing regulations to prevent environmental damage, and involving local communities in conservation efforts. By aligning tourism with environmental preservation, we can ensure the long-term viability of mountain destinations and safeguard their natural beauty.

Secondly, institutional support and capacity building are vital for the formalization and growth of tourism in conflict-affected mountain territories. Partnering with governmental and non-governmental organizations can provide mountain dwellers and tourism cooperatives with valuable resources, training, and expertise. This support should focus on improving knowledge and skills related to hospitality, guiding, environmental conservation, and business management. Additionally, strengthening regulatory frameworks and governance structures specific to mountain tourism can create a supportive environment that fosters responsible and sustainable tourism practices (Yang et al., 2009).

Future Research

Future research is urgently needed to fill in the gaps and examine the possible advantages of dark tourism for mountain people in areas plagued by armed conflict. The current mini-case study sheds light on a tourism cooperative established by FARC victims' women in Colombia's mountainous Chalán region, which is only one case study, though; more is required to fully comprehend how dark tourism contributes to other conflict-affected mountain communities' economic, social, and environmental well-being.

Future studies must focus on dark tourism's social and economic impacts on mountain residents, which can be assisted by looking at the perspectives and experiences of locals, particularly ex-combatants and war victims participating in dark tourism programs. By comprehending their perspectives, researchers can determine the precise ways in which dark tourism can aid in reducing poverty, empowering communities, and promoting sustainable development in mountainous locations. Investigating the ecological repercussions of violent conflicts in mountainous regions is crucial to developing environmental restoration and conservation project plans.

Future studies should also investigate dark tourism's governance and policy aspects in mountainous areas that have experienced war, which is part of assessing how local and national governing authorities can encourage sustainable practices and ensure that the benefits of tourism are equitably distributed to all stakeholders. Studying regulatory frameworks and policy interventions can also provide useful knowledge on controlling conflicts of interest, reducing the negative consequences of "dark tourism," and preserving mountainous regions' natural and cultural heritage.

Future research should generally focus on improving our comprehension of dark tourism as a viable tactic for benefiting the economic, social, and environmental interests of mountain locals, potential tourists, and the greater community, which can be achieved using case studies, interdisciplinary approaches, and collaboration between academics, policymakers, local communities, and tourism stakeholders. By filling in these knowledge gaps, we can provide valuable guidance for boosting mountain tourism practices, safeguarding the natural beauty of these locations, and improving the general well-being of local inhabitants.

The potential impact of mountaineering tourism on Colombia's natural environment, particularly in areas formerly controlled by guerrilla groups, is a subject ripe for research. These regions, once inaccessible due to conflict, are now opening up, revealing pristine natural environments that have been largely untouched and protected as a consequence of restricted access. The transition from conflict zones to tourist destinations poses both opportunities and challenges. On the one hand, there is the potential for economic development and community rehabilitation through sustainable tourism. On the other hand, there are concerns about the environmental impact and the need for sustainable practices to preserve these newly accessible ecosystems. The situation presents a valuable opportunity for research into sustainable tourism models that balance economic development with environmental conservation and community participation in regions emerging from conflict.

Despite these challenges, dark tourism has come to light as an alternative route for community growth and rehabilitating lives in mountainous regions ravaged by armed conflict. Dark Tourism projects seek to create economic opportunities, encourage peace, and improve mountain dwellers' general well-being by utilizing these areas' unique natural and cultural resources (Wengel et al., 2021).

Management Implications

Based on the insights gathered from this study, several management implications can be outlined to guide other mountainous regions globally. First, there is a pressing need to integrate dark tourism within the broader spectrum of sustainable mountain tourism development. This necessitates a balanced approach that respects the delicate ecological, cultural, and historical contexts of such regions. Managers and policymakers should prioritize the preservation of natural landscapes and cultural heritage, ensuring that tourism development does not compromise these assets.

To harness the potential of dark tourism for local development, it is essential to involve community members in the planning and management processes. This inclusion fosters a sense of ownership and responsibility among the local population, leading to more sustainable and ethically managed tourism activities. Training programs aimed at enhancing local capacities in hospitality, tour guiding, environmental conservation, and cultural interpretation can significantly contribute to creating meaningful and enriching visitor experiences, while simultaneously boosting local economies.

Moreover, the development of dark tourism should be aligned with broader peacebuilding and reconciliation efforts in post-conflict settings. By narrating the stories of conflict and recovery through guided tours and interpretative materials, dark tourism can play a pivotal role in educating visitors about the historical and socio-political dynamics of the region. This educational aspect can promote understanding, empathy, and healing, both for visitors and local communities affected by past conflicts.

Strategically marketing these unique tourism offerings to global audiences, while ensuring that narratives are respectful and authentic, can attract a diverse visitor base interested in understanding the complex layers of history and culture inherent in these mountainous regions. Collaborative marketing efforts, leveraging digital platforms and partnering with international tourism agencies, can help in reaching wider markets effectively.

Finally, it is crucial to monitor and evaluate the impact of dark tourism initiatives continuously. This should include assessing environmental sustainability, visitor satisfaction, community benefits, and cultural sensitivity. Such evaluations can inform adaptive management strategies, ensuring that dark tourism contributes positively to the holistic development of mountainous regions, aligning with global best practices in sustainable and responsible tourism.

Funding

Part of this research has been partially supported by the Colombian Ministry of Science, Technology, and Innovation [grant number 909–2021] and by the doctoral program in management of organizations from the International School of Economic and Administrative Sciences at Universidad de La Sabana in Colombia and the Universidad de Salamanca (Spain) (Grant Ayudas María

Zambrano para la atracción de Talento Internacional funded by MCIN/AEI/10.13039/501100011033 by "European Union NextGenerationEU/PRTR").

References

Apollo, M. (2017). The true accessibility of mountaineering: The case of the High Himalaya. *Journal of Outdoor Recreation and Tourism*, 17, 29–43. https://doi.org/10.1016/j.jort.2016.12.001.

Apollo, M., & Rettinger, R. (2019). Mountaineering in Cuba: improvement of true accessibility as an opportunity for regional development of communities outside the tourism enclaves. *Current Issues in Tourism*, 22(15), 1797–1804. https://doi.org/10.1080/13683500.2018.1446920.

Apollo, M., & Wengel, Y. (2022). *Mountaineering Tourism. A Critical Perspective*. https://doi.org/10.1007/978-4-431-99659-0_54.

Bowman, M. S., & Pezzullo, P. C. (2009). What's so "Dark" about "Dark Tourism"?: Death, Tours, and Performance. *Tourist Studies Performance*, 9(187). https://doi.org/10.1177/1468797610382699.

Cantillo, T., & Garza, N. (2022). Armed conflict, institutions and deforestation: A dynamic spatiotemporal analysis of Colombia2000–2018. *World Development*, 160, 106041. https://doi.org/10.1016/j.worlddev.2022.106041.

Chen, S., Wang, S., & Xu, H. (2017). Influence of Place Identity on Residents' Attitudes to Dark Tourism. *Journal of China Tourism Research*, 13(4), 338–356. https://doi.org/10.1080/19388160.2017.1401023.

Cornell, D. A. V., Tugade, L. O., & Perez, A. J. (2019). The bright side of dark tourism in Baguio city: Understanding its impact to the stakeholders. *Journal of Tourism and Development*, 31, 61–76.

Corredor-Garcia, J., & López, F. (2023). The Logic of " War on Deforestation ": A Military Response to Climate Change in the Colombian Amazon. *Special Issue: – Militarism and Militarization in Latin America. Alternatives: Global, Local, Political*, 0(0), 1–19. https://doi.org/10.1177/03043754231181741.

Frew, E. A. (2012). Interpretation of a sensitive heritage site: The Port Arthur Memorial Garden, Tasmania. *International Journal of Heritage Studies*, 18(1), 33–48. https://doi.org/10.1080/13527258.2011.603908.

Garavaglia, V., & Diolaiuti, G. (2012). Evaluating Tourist Perception of Environmental Changes as a Contribution to Managing Natural Resources in Glacierized Areas: A Case Study of the Forni Glacier (Stelvio National Park, Italian Alps). *Environmental Management*, 50, 1125–1138. https://doi.org/10.1007/s00267-012-9948-9.

Government of Sucre (2022). Gobernación de Sucre: Departamento de Sucre. https://www.sucre.gov.co.

Kamber, M., Karafotias, T., & Tsitoura, T. (2016). Dark heritage tourism and the Sarajevo siege. *Journal of Tourism and Cultural Change*, 14(3), 255–269.

Kumar, S., Sureka, R., & Vashishtha, A. (2020). The Journal of Heritage Tourism: a bibliometric overview since its inception. *Journal of Heritage Tourism*, 15(4), 365–380. https://doi.org/10.1080/1743873X.2020.1754423.

Lemelin, H., Dawson, J., Stewart, E. J., Maher, P., & Lueck, M. (2010). Last-chance tourism: The boom, doom, and gloom of visiting vanishing destinations. *Current Issues in Tourism*, 13(5), 477–493. https://doi.org/10.1080/13683500903406367.

Light, D., Rugg, J., & Young, C. (2023). The disposal of cremation ashes in tourism settings: practices, impacts and management. *Current Issues in Tourism*, 26(8), 1354–1366. https://doi.org/10.1080/13683500.2022.2054403.

Mansfeld, Y., & Korman, T. (2015). Between war and peace: Conflict heritage tourism along three Israeli border areas. *Tourism Geographies*, 17(3), 437–460. https://doi.org/10.1080/14616688.2015.1036916.

Musa, G., Hall, C. M., & Higham, J. E. S. (2004). Tourism sustainability and health impacts in high altitude adventure, cultural and ecotourism destinations: A case study of Nepal's Sagarmatha national park. *Journal of Sustainable Tourism*, 12(4), 306–331. https://doi.org/10.1080/09669580408667240.

Nepal, S., & Chipeniuk, R. (2006). Mountain Tourism: Toward a Conceptual Framework Mountain Tourism. *Tourism Geographies*, 7(3), 37–41. https://doi.org/10.1080/14616680500164849.

Olson, K. R., & Speidel, D. R. (2023). United States Secret War in Laos: Long-Term Environmental Impacts of the Use of Chemical Weapons. *Open Journal of Soil Science*, 13, 199–232. https://doi.org/10.4236/ojss.2023.134009.

Ospina, G. A. (2023). Mountains in the Context of War in Colombia: Some Geopolitical Considerations. International Mountain Society. *Mountain Research and Development*, 26(1), 90–91. https://doi.org/10.1659/0276-4741(2006)026[0090:MITCOW]2.0.CO;2.

Palomo, I. (2017). Climate Change Impacts on Ecosystem Services in High Mountain Areas: A Literature Review. *Mountain Research and Development*, 37(2), 179–187. https://doi.org/10.1659/MRD-JOURNAL-D-16-00110.1.

Prahalad, & Hart, S. (2010). The fortune at the bottom of the pyramid. *Revista Eletrônica de Estratégia & Negócios*, 1(2), 1. https://doi.org/10.19177/reen.v1e220081-23.

Prayag, G., Suntikul, W., & Agyeiwaah, E. (2018). Domestic tourists to Elmina Castle, Ghana: Motivation, tourism impacts, place attachment, and satisfaction. *Journal of Sustainable Tourism*, 26(12), 2053–2070. https://doi.org/10.1080/09669582.2018.1529769.

Qian, L., Zhang, J., Zhang, H., & Zheng, C. (2017). Hit close to home: the moderating effects of past experiences on tourists' on-site experiences and behavioral intention in post-earthquake site. *Asia Pacific Journal of Tourism Research*, 22(9), 936–950. https://doi.org/10.1080/10941665.2017.1362019.

Qian, L., Zheng, C., Wang, J., Pérez Sánchez, M. D. L. Á., Parra López, E., & Li, H. (2022). Dark tourism destinations: the relationships between tourists' on-site experience, destination image and behavioural intention. *Tourism Review*, 77(2), 607–621. https://doi.org/10.1108/TR-08-2020-0360.

Saini, G. K., Lievens, F., & Srivastava, M. (2022). Employer and internal branding research: a bibliometric analysis of 25 years. *Journal of Product and Brand Management*, 31(8), 1196–1221. https://doi.org/10.1108/JPBM-06-2021-3526.

Sayed, A., Hashimy, Q., & Law, P. D. S. (2023). Impact of War on the Environment: A critical study of Afghanistan. *International Environmental Legal Research Journal*, I (I), 52–61. http://dx.doi.org/10.2139/ssrn.4352538.

Skinner, J. (2018). Plymouth, Montserrat: apocalyptic dark tourism at the Pompeii of the Caribbean. *International Journal of Tourism Cities*, 4(1), 123–139. https://doi.org/10.1108/IJTC-08-2017-0040.

Stefano, D., & Letey, M. (2019). The role of a national park in classifying mountain tourism destinations: An exploratory study of the Italian Western Alps. *Journal of Mountain Science*, 16, 1675–1690. https://doi.org/0.1007/s11629-018-5356-9.

Stone, P., & Grebenar, A. (2022). "Making Tragic Places": dark tourism, kitsch and the commodification of atrocity. *Journal of Tourism and Cultural Change*, 20(4), 457–474. https://doi.org/10.1080/14766825.2021.1960852.

Stone, P., & Sharpley, R. (2008). Consuming Dark Tourism: A Thanatological Perspective. *Annals of Tourism Research*, 35(2), 574–595. https://doi.org/10.1016/j.annals.2008.02.003.

Wang, J., Wang, G., Zhang, J., & Wang, X. (2021). Interpreting disaster: How interpretation types predict tourist satisfaction and loyalty to dark tourism sites. *Journal of Destination Marketing and Management*, 22. https://doi.org/10.1016/j.jdmm.2021.100656.

Wengel, Y., Aye, N., Yee, W., Pyar, K., & Kreisz, J. (2021). *Mountainous Protected Areas in Myanmar: Current Conditions and the Outlook for Nature-Based Tourism.* Springer. https://doi.org/10.1007/978-3-030-76833-1.

Witte, A., Wilson, J., Burrai, E., Dashper, K., Witte, A., Wilson, J., Burrai, E., Navigating, K. D., Witte, A., Wilson, J., & Burrai, E. (2023). Current Issues in Tourism Navigating tourism ethnographies – fieldwork embroiled in time, movement and emotion movement and emotion. *Current Issues in Tourism*, 26(9), 1394–1408. https://doi.org/10.1080/13683500.2022.2057841.

Wu, H.-C., & Cheng, C.-C. (2018). What drives supportive intentions towards a dark tourism site? *International Journal of Tourism Research*, 20(4), 458–474. https://doi.org/10.1002/jtr.2196.

Yang, M., Hens, L., Ou, X., & De Wulf, R. (2009). Tourism: An alternative to development?: Reconsidering farming, tourism, and conservation incentives in northwest Yunnan mountain communities. *Mountain Research and Development*, 29(1), 75–81. https://doi.org/10.1659/mrd.1051.

7 Pro-Poor Mountain Tourism Employment for Women in Nepal

Wendy Hillman

Introduction

Researchers have demonstrated that there is an underlying connection concerning tourism development and economic development, and between poverty reduction and tourism (Zhao & Ritchie, 2007; Ashley & Mitchell, 2009; Tucker & Boonabaana, 2012; Croes & Vanegas, 2008). It has been named "pro-poor tourism". Pro-poor tourism is also labelled as the "ethical cousin of fair trade in tourism" (Seif, 2004). Performing inclusive tourism promotes relationships involving diverse facets of the tourist sector. For example, collaborations with specific contributors, assimilation of women and participation of rural populations.

Therefore, this sees tourism as carrying out a primary function in the struggle against poverty. Tourism's influence towards poverty reduction, and thus, towards pro-poor tourism, has expanded appreciably. Tourism enriches the relationship linking the tourism industry and poor populations so that they are able to partake more successfully in tourism innovation and development (Bolwell & Weinz, 2008; Sinha & Rishad, 2016).

Pandey (2008), argues that the addition of the poor section of the community in all types of community-based tourism, incorporating pro-poor mountain tourism, is a kind of tourism activity that supports environmental, social, and cultural conservation. Pro-poor mountain tourism is directed by the community and managed to assist tourists to build-up their knowledge and gain an understanding of the people so that their participation ensures benefit to the population. Mountain tourism affords a possibility of utilising local wisdom for the socioeconomic advancement of impoverished populations in the mountains (including women), conditional on making it ecologically viable and pro-poor (Upadhayaya & Upreti, 2011).

Background

In developing economies, such as Nepal, women's labour both inside and outside the family sphere, such as in the informal sector and in food production, is so underestimated that their participation in community and

DOI: 10.4324/9781003392583-8

nationwide progress continues to be mostly unseen, unillustrated in nationwide indicators and forgotten in national decision formulation. In comparison to their percentage of the population, women are not considered as equivalent contributors in development or in pursuits and programs of adult and informal training (Bernard & Gayfer, 1983).

Traditionally in Nepal, girls marry early and are restricted to the domestic sphere. Employment external to the household and securing wages to sustain their family unit are not fostered. One way to overcome this situation, and to warrant the presence of women in pro-poor tourism is to encourage women-owned initiatives. This, in turn, can encourage women's economic self-determination, assist them in overcoming their shortage of resources, energise them, and supply them with participation in their communities. In Nepal's Constitution, both women and Dalits are acknowledged as ostracised groups. Ostracised groups are those with a reduced amount of admittance to civil, financial and societal prospects because of their caste, cultural group or gender. Dalits are the de facto "untouchables" of contemporary Nepal. Dalit refers to a group of people who are religiously, culturally, socially, economically and historically oppressed, excluded and treated as untouchables and they belong to different geographical region, language, culture and castes (Dalit Welfare Organization, 2018). Females who were formerly reliant on their spouses for financial capital are now able to supplement the household earnings (Adhikari, et al., 2018) through engagement with pro-poor tourism (see Figure 7.1).

The literature on tourism development includes many instances of structures and ideas devised to control tourism effects for the gain of tourists, local communities and nature (Richards & Hall, 2000). Beginning in the 1950s

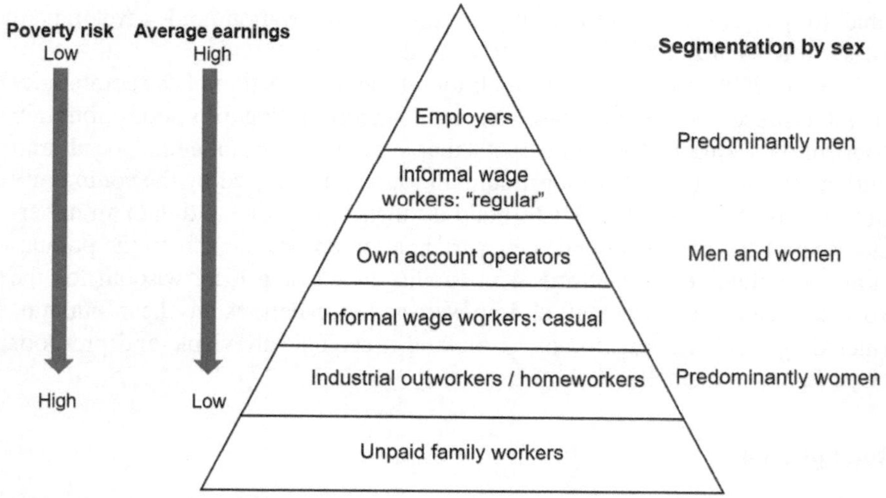

Figure 7.1 Women in informal employment: Globalizing and Organizing (WIEGO) model of informal employment – hierarchy of earnings and poverty risk (Source: adapted from Chen, 2012)

and 1960s (which coincided with the opening of the Nepali borders to tourism for the first time – see Nepal, 2022), travel was encouraged as a process through which "Third World" nation-states could receive foreign exchange, generate work and "develop" their financial system similar to western society (Hummel & van der Duim, 2012; Spenceley & Meyer, 2012), in keeping with the reasoning of modernisation theory (Scheyvens, 2009). Disheartened with the inefficient "trickle-down" trade and industry of wholesale sightseeing (Scheyvens, 2009), travel market analysts in the 1970s adjusted their attention to community-based tourism (CBT) attitudes which highlighted modest and locally held enterprises (Hummel & van der Duim, 2012; Tosun & Jenkins, 1998). This adjustment in attention led to a diversity of "alternative tourism" perceptions, the most lasting of which have been "ecotourism" and "sustainable tourism" (Espiner, Stewart & Lama, 2017); areas where women have stood out (Hillman & Radel, 2022a).

Even though the poverty-stricken tend to carry the problem of the damaging ecological effects that tourism at times triggers, poor females commonly suffer considerably more than males. For instance, in the Mid-Hills area of Nepal "deforestation linked to fuel for tourism made it harder for village women to collect wood fodder and leaves for mulching and manure" (Shah & Gupta, 2000, p. 38). On the contrary, sexual abuse in tourism involves both males and females (Shah & Gupta, 2000, p. 39). All of these challenges must be dealt with in order for tourism to grow to be pro-poor. Clearly, in many circumstances and situations, tourism will not be a suitable approach for poverty decrease, whether as a result of socio-acculturation, ecological preservation, occupation, or other causes (Cattarinich, 2001).

Adshead claims that, in Nepal, social norms determine that nearly all of the vacancies and casual work that are accessible in tourism (i.e., as guides and porters) ought to be given to males, even though one tour operator does demand that in-bound tourism operators ought to employ females when consumer groups are comprised mainly of females. The occupation business spheres established there (see Apollo & Wengel, 2022; Hillman, 2022; Hillman, 2019a; Hillman, 2019b; Hillman & Radel, 2022a; Hillman & Radel, 2022b; Jackson, 2010).

Numerous females operate small-scale guesthouses (also called teahouses) and dining halls in the trendy trekking destinations of Annapurna, Everest, and Langtang. Undeniably, more females than males work in the lodging sphere in those areas (Shah & Gupta, 2000). In spite of that, across the nation only 1 in 8 women work in the lodging area, and consistent with Shrestha and Shrestha (2021), in 2019 only 23 of 185 of those working in the safari retreats of Chitwan National Park were females. Thus, 78% were males and 12% were females (Shrestha & Shrestha, 2021).

The Annapurna Conservation Area Project manages the Developing Women's Entrepreneurship in Tourism (DWET) program, which offers flexible loans to females to initiate added income producing undertakings (Shah & Gupta, 2000, p. 38). Shah and Gupta (2000) have identified that few studies consider the issues that may affect female employment styles in Nepal, "such as comparison with the employment of women in other organised sectors, the availability of trained personnel, cultural traditions or sexual discrimination" (p. 38). The experience of Muir's Tours implies that cultural traditions in part establish the sphere of the industry where females can play a part in addition to the degree to which they will contribute. While Sherpa, as a profession, shape the Everest region, a blend of cultural groupings inhabit the Annapurna and Langtang treks, and also the Terai (Cattarinich, 2001).

Findings

From a pro-poor tourism point of view, a small number of misgivings are identifiable within this research subject. There is a good deal of disparity in the scope in which females gain from pro-poor tourism schemes. In conditions of employment or informal work, females and males seemingly gain almost the same levels of equivalence from communal pro-poor tourism enterprises in Prainha do Canto Verde and from Ecotour Samoa's enterprises. Nonetheless, females embody less than 20% of the underprivileged who gain from the informal employment possibilities that were initiated by IntoAfrica in Kenya and Tanzania. Maurice Adshead recounts that females in Nepal are also less prone than males to gain from informal employment. Notwithstanding, prior inquiry in Nepal implies that females from certain cultural groups and in specific areas (i.e., Sherpa females in the Khumbu Himal) are effective at operating teahouses and eating establishments (Cattarinich, 2001).

Nepal is one of the peak prospective tourism nations globally. The government has recognised that environmental tourism expansion is an important means for poverty lessening across the nation. Kafle (2011) has established that tourism is acknowledged as the most substantial form of commerce in Nepal which produces multi-administrative, multidimensional and multi-comprised prospects for poverty lessening. Nepal reaps the relative rewards over earlier developed economies with reference to societal tradition, ecological assets, environmental range and time-honoured pursuits without hi-technological knowledge and workforce-concentrated industry. The dispersed character of tourism in Nepal – natural surroundings/eco-diversity, exploration, customs, ancient, spiritual and other sectors echo the ability of knowing the environmental tourism challenge for key accommodating sectors. On the subject of being a developing maintainable sector, Nepal needs to progress environmental tourism expansion in addition to promotion stratagems to influence likely tourists and market sectors (Kafle, 2011) within pro-poor mountain tourism.

In September 2001, the Government of Nepal (GoN) initiated The Tourism for Rural Poverty Alleviation Program (TRPAP). This program specifically concentrated on the strategies regarding pro-communal, pro-female, pro-poverty, and pro-ecosystem. The specialised and economic support was made available by the United Nations Development Program (UNDP)/Nepal, Department for International Development (DFID)/ Nepal and SNV/Nepal (SNV Netherlands Development Organisation). The scheme was planned to be entrusted to the Nepal Tourism Board (NTB) after completion, with the formation of a Sustainable Tourism Development Unit (STDU) on the Board (Kafle, 2011).

Nepal has followed environmental tourism as a device of tourism sector development in addition to the social and economic growth of Nepal; which has also enhanced pro-poor tourism in recent times. The Ministry of Culture Tourism and Civil Aviation (MoTCA) and the NTB have prepared a beneficial strategy, guidelines and policies; progressing associated vital systems; augmenting the abilities of workers; joining forces with various interested parties; endorsing and enabling the expansion of environmental tourism from government ranks. As a nationwide aspiration, the TRPAP and additional activities initiated by the STDU of the NTB in distinctive topographical locations in tourism localities are pursuing similar activities such as the success of addressing pro-poor tourism in Nepal. Application and extension of TRPAP suggests an accord linking environmental tourism growth, advertising stratagem and poverty lessening in Nepal. Several pointers established by using this approach have been observed to have a constructive influence on tourism for poverty lessening (Kafle, 2011).

Tourism expansion actions at the current time in Nepal have meaningfully assisted the poor across tourism spheres. Most tourism areas in Nepal have shifted from poverty-based towards a more standardised lifestyle, including in the Himalaya areas and remote regions. The procedure of lessening poverty in these regions is more effective in contrast with other impoverished groups of the non-tourism areas. The local, pastoral, biological, community, athletics, mountain climbing, hiking, familiarisation tours, bird surveillance, jungle expedition, rafting, training, acculturation, agri- and wellness tourism are to a certain degree meeting the environmental tourism need for key producing sectors (Kafle, 2011).

Nearly all characteristics of alternative and pro-poor tourism have concentrated on the conservation of resources and gains accumulating for populations in specific areas (Ashley, Boyd & Goodwin, 2000). Yet, several of these characteristics have not explicitly respected the requirements and wishes of the most underprivileged people of host populations, an inclusion somewhat confronted by the comparatively contemporary development of the pro-poor-tourism stance (Ashley et al., 2000). Pro-poor tourism's focus is on "unlocking opportunities for the poor within tourism, rather than expanding the overall size of the sector" (Goodwin,], 1999, p. 6). It is an attitude to the development and supply of tourism that creates overall financial, ecological, communal or traditional advantages for the poor by augmenting their involvement through skills building and knowledge transfer (Espiner, Stewart & Lama, 2017; Ashley, Roe, & Goodwin, 2001a; Goodwin, 1999).

In general, the notion of pro-poor tourism is similar to other tourism development types, such as cultural-based tourism, ecotourism and sustainable tourism, all of which seem to encourage broad, community-level and collectively accountable attitudes towards tourism progress. As Chok, Macbeth, and Warren (2007) have stated, the pro-poor tourism dialogue, similar to that of environmental improvement, is not governmentally neutral. While the expansive intentions and overall agenda have a broad influence, there is contested agreement over completion and capacity of results (Harrison, 2008). Nonetheless, consistent with Ashley et al. (2001a, p. 2); in spite of its "triple bottom line" ideologies, sustainable tourism has appeared to underline ecological maintainability, as opposed to the cultural, economic and social features of pro-poor tourism. For instance, the primary focus of Agenda 21 for the tourism and travel industry is on ecological maintainability (Espiner, Stewart & Lama, 2017).

Correspondingly, ecotourism is frequently conscientiously and firmly concentrated on preservation of the local community ecosystems. Whereas cultural-based tourism seeks out the engagement of populations in tourism development and planning, ecotourism characteristically considers the population as uniform, not distinguishing between likely hierarchies in each population (Ashley et al., 2001a). Pro-poor tourism incorporates qualities of the ideas already discussed here, but limits the societal emphasis significantly, necessitating detailed procedures to afford possibilities for the most impoverished populations to participate at all stages and tiers of tourism management (Goodwin, 1999). Chok et al. (2007) imply that pro-poor tourism inhabits a fragile environmental /strong theoretical locus, acknowledging that ecological advantages are subordinate to gains ensuing to impoverished communities (Espiner, Stewart & Lama, 2017).

Discussion

Despite the fact that pro-poor tourism has extensive intrinsic attraction and, as Harrison (2008, p. 858) has stated, "[i]t seems so evidently morally correct … [-] after all, who could oppose the interests of the poor?"; however, the model is not without its critics. Notable among these critiques are the assertions that pro-poor tourism is theoretically vague or unclear and that its effects have not been, or cannot be, precisely determined. Harrison (2008), for example, surmises that pro-poor tourism neglects to distinguish itself as a method to tourism growth, requiring a constant focal point; and minus distinct techniques, instead outlining it as a "movement" or "pressure group" with a stated concern for the poor. Others, such as Chok et al. (2007), have argued that pro-poor tourism is "morally charged", subject to rhetoric and vulnerable to political manipulation. A further criticism of the pro-poor tourism approach embodies the idea that pro-poor tourism does little to overcome the fundamental barriers to redistribution of tourism's benefits, and in many cases fails to reach the very poorest members of the target communities (Espiner, Stewart & Lama, 2017; Harrison, 2008; Scheyvens, 2007). In this case, women in Nepal.

Mindful of this result are a collection of structural circumstances, social, political and economic in nature that more often than not eliminate some people from effective involvement in the economy. Likened to earlier tourism development methods, pro-poor tourism does not typically represent a "new start" in practice, guiding Chok et al. (2007, p. 155) to distinguish an " … almost schizophrenic quality to the pro-poor tourism discourse, as it upholds moral arguments for poverty alleviation within a predominantly business-as-usual environment". Spenceley and Meyer (2012) and Scheyvens (2007) have also argued that, as a substitute for refuting the dominant neoliberal ideology, pro-poor tourism complements and extends it: " … some organisations are adopting PPT rhetoric primarily as a means of promoting trade liberalisation, market-led growth and private sector development" (Espiner, Stewart & Lama, 2017; Scheyvens, 2007, p. 247).

While not wholly set aside for pro-poor tourism, an additional frequent account in tourism growth texts concerns the restrained confirmation of advantages accumulating to poor or forgotten populations within tourism. Whereas many writers recognise the "potential" of tourism growth to better circumstances or augment means of support options for regional and remote underprivileged populations in general (Bello, Carr, & Lovelock, 2016; Goodwin, , 1999; Goodwin, 2008, 2009; Harrison, 2008; Scheyvens, 2007; Spenceley & Meyer, 2012; Zeng & Ryan, 2012), there is an absence of conclusive realistic confirmation of tourism's influence to the mitigation of poverty in developing economy circumstances (Chok et al., 2007; Hummel & van der Duim, 2012).

The "pro-poor" emphasis over approximately the last 10 years has yet to invert this tendency and although, possibly as an outcome of the pro-poor tourism campaign, destitution and privation may well have a heightened profile, measurable instances of the pro-poor affects are sporadic (Harrison, 2008; Hummel & van der Duim, 2012) at best. However, it is essential to recognise, that uninterrupted financial processes are only one section of growth illustration, and pro-poor undertakings related to tourism progress could also be counted as positive with regard to basic organisational advances (Lemelin, Koster, & Youroukos, 2015), or advances in self-confidence and community emancipation (Bennett, Lemelin, Koster, & Budke, 2012; Scheyvens, 1999) through employment in the tourism development sector (Espiner, Stewart & Lama, 2017; Bith, Espiner, & Cushman, 2011).

Despite these theoretical and fundamental criticisms, the pro-poor tourism idea does offer a suitable structure within which to observe processes amidst a distinct representation of involved community populations who take part in tourism development activity. It is openly asserted elsewhere in the literature (Chok et al., 2007; Goodwin, 2009; Harrison, 2008; Zeng & Ryan, 2012), that the arrival of pro-poor tourism in tourism growth terminology has brought about an identifiable dilemma to those populations on the borders of financial pursuit and has also necessitated an examination of tourism expansion standards and procedures. Of special importance is the weight that advocates of pro-poor tourism use regarding contributory planning procedures that sustain and facilitate capability development in resident communities in the course of pro-poor tourism activities (Espiner, Stewart & Lama, 2017; Bennett & Lemelin, 2014; Saufi, O'Brien, & Wilkins, 2014).

Recognising both the possibility and drawbacks of tourism is a basic primary phase regarding rendering the industry more pro-poor. However, what is lacking from some micro-stage inquiries, is an evaluation of the countless explicit and implicit effects that tourism may have in a specific locale and how they link to the identifiable occupation ideas of the local communities. It is one function to recognise "positive impacts"; but it is another to reflect on exactly how they correspond to the occupations of the poverty-stricken, in relation to pro-poor tourism philosophies. For instance, one of the drawbacks is that most tourism occupations for resident individuals in developing economies are seasonal. This position also accepts that permanent salary employment is the most appropriate method of financial pursuit. Nevertheless, this opinion does not include explanation of how income occupation corresponds with prevailing income behaviours of the local inhabitants. Correspondingly, a blend of diverse income possibilities lessens the threat of considerable reliance on what can be an unpredictable business (Cattarinich, 2001).

Nonetheless, in suitable circumstances the point that tourism rarely has capitalised on its pro-poor capability suggests that there is sizeable room for refining the configuration of this area of tourism and augmenting the prospects that the underprivileged may gain from it, some of whom are currently contributing to the industry. As Brohman (1996) remarks:

> *… if stress is not placed on the creation of local linkages to spread the benefits of growth in social, sectoral, and regional terms, neoliberal outward-oriented strategies risk replicating the vicious cycles of polarisation and repression so commonly associated with past export-oriented development models. What is missing from strategies that focus only on increasing international trade or tourism is a concern for the broader development goals of raising living standards of the popular majority and promoting more balanced growth among different economic sectors and geographic regions. In the absence of well-developed linkages between the external sectors and the rest of the economy, a limited and polarised form of development takes place that cannot act as a stimulus for broadly based development.*
>
> (Cattarinich, 2001, p.50)

With the purpose of maximising tourism's poverty lessening capacity, various approaches may be needed that blend action in the tourism location, at a nationwide policy level, and also globally. Because one organisation cannot function efficiently at all three levels, corresponding activities by various interested parties are needed at all levels to increase the affirmative results of pro-poor tourism. The destination point depends predominately upon programs by private businesses, NGOs, and underprivileged populations themselves. Nationwide administrations can decrease difficulties for unofficial sector involvement. For instance, governments can safeguard that location guidelines allow resident entrepreneurs, and especially females, entry to tourism spheres (see Apollo & Wengel, 2022). They can also augment the resources of the underprivileged on which tourism commodities can be advanced through, for example, land freehold and scholastic programs, in addition to increasing their engagement in decision producing procedures that impact their sources of revenue (Goodwin, 1999).

An obvious oversight in the literature is an investigation of the particular activities that NGOs, district groups, individual businesses, and governments are carrying out to prompt tourism to be more pro-poor, and the influences that these projects have had on the incomes of the underprivileged. Once more, it is one thing to maintain what ought and ought not be achieved to make tourism pro-poor, but it is a completely different issue to gauge the effectiveness of specific programs undertaken by particular individuals in specific socio-cultural, economic, political, and environmental circumstances (Cattarinich, 2001).

Conclusion

Pro-poor tourism can create change for the underprivileged and their existences and can "tilt" the mountains in the direction of the poor, oftentimes only minimally. Integrating an emphasis on poverty throughout the tourism sector, particularly in the mountains could be seen as a daunting task. On the other hand, in view of the significance of tourism in various disadvantaged regions, unquestionably, it is worthwhile pursuing this problem (Roe, Goodwin & Ashley, 2004).

Pro-poor tourism contributes to adversity lessening. In spite of the decline in tourist figures to the Karnali Zone, far north-western Nepal, community mobilisation, tourism mindfulness and commercial backing assistance have brought about minor-level endeavours concentrating on provincial spheres and visitors, dispersing pressures for needier business operators. Pro-poor mountain tourism can be established in districts that have a tourism capacity with marketable feasibility. These type of programs call for a strong pro-poor emphasis. This should incorporate options for involvement of the poor, founded on tourism understanding, as impoverished populations frequently perceive tourism as a means to ameliorate their income, but do not have the necessary expertise and skilfulness; and the establishment of financial growth backing. Right of entry of the impoverished to tourist spheres persists as an significant issue (locality of the sector, financial leaders, societal restraints on underprivileged enterprises). All these components should be considered if pro-poor tourism programs are to have an authentic impact on adversity lessening. As is clear from the example of the Karnali Zone, peripheral issues can also exercise a significant effect. Pro-poor mountain tourism has to take a sector-inclusive direction, operating at the local, national and universal levels. This is central to the purpose of maximising cooperation between supporters encouraging formal progress at the community level, but also as an influence on nationwide strategies (Hummel, 2004, in Luger, Baumgartner & Wöhler, 2004).

Considering 85% of the overall population and 95% of the underprivileged live in isolated areas, poverty in Nepal is a rural experience. Additionally, with more than two-thirds of the inhabitants directly active in the food production area, development in Nepal will be pro-poor if it takes place in isolated areas and improves the output of the agronomy sector. The impoverished need added assistance to escape the poverty web. A secure approach that highlights rural area development, community financing in infrastructure that employs the provincial underprivileged; and allows for the adverse effects of a restrained fiscal policy should be understood as a part of a pro-poor tourism growth agenda (Shrestha, 1999).

So ultimately, if women do indeed "hold up half the sky", then the plight of Nepali women and pro-poor tourism in Nepal can only, over time, contribute to a stronger pro-poor tourism base, in the country; and particularly in the mountains, and lead the way to contribute to growth, prosperity and a pathway away from poverty via tourism and the tourist sector.

Management Implications

The role of government, private sector players, NGOs, community organisations, and the people themselves within pro-poor mountain tourism is critical and multifaceted. The private sector can directly contribute to pro-poor partnerships. Private operators should, at the very least, be involved in the development of products and markets to ensure commercial reality. There are many things that only the government can do, and a prominent role for the government in the process of poverty alleviation is a great benefit. At the very least, there must be a policy framework that facilitates poverty alleviation. Poor people themselves are essential for poverty alleviation, but they might need to be organised at community level to play an effective role in tourism (Ashley, et al., 2001b).

Having an additional organisation to catalyse and support others' pro-poor mountain tourism efforts is often, but not always, an NGO's role. Donors can also support pro-poor mountain tourism through their role in support of tourism plans and the "sustainable tourism" agenda. Poverty reduction through pro-poor mountain tourism has the potential to be significant at local or district levels. Pro-poor tourism strategies do seem capable of "tipping" the industry at the margins to open up opportunities for the poor, and they have the potential to be broadly applied across the industry. So far, however, pro-poor mountain tourism initiatives have made little impact on national aggregates, as initiatives are small scale, site specific, or at an early stage of implementation. National impacts would necessitate a shift across the industry and would vary depending on the location and relative size of the tourism industry (Ashley, et al., 2001b).

Even so, if all of the world's major tourist destinations were accessible to the poor, it would impact millions of them. How cost-effective is pro-poor mountain tourism? Who can say: pro-poor mountain tourism initiatives have been "learn-by-doing", and they have taken a lot of time and money to implement. But if future local income, non-monetary benefits, and policy changes for the poor were taken into account, a "cost-benefit" approach would likely result in a positive return on investment. Sharing of experiences would help reduce the cost of implementing pro-poor mountain tourism approaches (Ashley, et al., 2001b).

References

Adhikari, L., Shrestha, A. J., Dorji, T., Lemke, E., & Subedee, B. R. (2018). Transforming the lives of mountain women through the Himalayan nettle value chain: a case study from Darchula, far west Nepal. *Mountain Research and Development*, 38(1), 4–13.

Apollo, M. & Wengel, Y. (2022). *Mountaineering Tourism: A Critical perspective.* Routledge. https://doi.org/10.4324/9781003095323.

Ashley, C. & Mitchell, J. (2009). *Tourism and poverty reduction: Pathways to prosperity.* Routledge. https://doi.org/10.4324/9781849774635.

Ashley, C., Roe, D., & Goodwin, H. (2001a). *Pro-poor tourism strategies: Making tourism work for the poor (Pro-poor tourism report 1).* Overseas Development Institute. Retrieved from http://www.odi.org.uk/resources/download/2358.pdf.

Ashley, C., Roe, D., & Goodwin, H. (2001b). Pro-Poor Tourism Report No. *Pro-Poor Tourism Strategies: making tourism work for the poor, a review of experience*, The Russell Press, Nottingham, UK.

Ashley, C., Boyd, C., & Goodwin, H. (2000). *Pro-poor tourism: Putting poverty at the heart of the tourism agenda.* Natural Resource Perspectives, 51.

Bello, F.G., Carr, N., & Lovelock, B. (2016). Community participation framework for protected area-based tourism planning. *Tourism Planning & Development.* doi:10.1080/21568316.2015.1136838.

Bennett, N., & Lemelin, R. H. (2014). Situating the eco-social economy: Conservation initiatives and environmental organizations as catalysts for social and economic development. *Community Development Journal*, 49(1), 69–84. https://www.jstor.org/stable/26166138.

Bennett, N., Lemelin, R. H., Koster, R., & Budke, I. (2012). A capitals framework for appraising and building capacity for tourism development in aboriginal protected area gateway communities. *Tourism Management*, 33(4), 752–766. https://doi.org/10.1016/j.tourman.2011.08.009.

Bernard, A. & Gayfer, M. (1983). *Women Hold Up More Than Half the Sky.* A Third World Perspective on Women and Nonformal Education for Development. International Council for Adult Education, Toronto (Ontario), Canada.

Bith, B., Espiner, S., & Cushman, G. (2011, April 27–30). Ecotourism, empowerment and community development: The case of Yeak Laom Protected Area, Cambodia. In E. Wickens & J. Tilbury (Eds.), *Proceedings of the international conference on special interest tourism and destination management* (39–51). Buckinghamshire New University.

Bolwell, D., & Weinz, W. (2008). *Reducing poverty through tourism* (No. 424657). International Labour Organization.

Brohman, J.(1996). New directions in tourism for Third World development. *Annals of Tourism Research* 23(1), 48–70. https://doi.org/10.1016/0160-7383(95)00043-7.

Cattarinich, X. (2001). *Pro-poor tourism initiatives in developing countries: Analysis of secondary case studies.* PPT Working Paper No. 8, http://www.propoortourism.org.uk, Department of Sociology, University of Alberta, Edmonton, Alberta, T6G 2H4, CANADA, T6G 2H4.

Chen, M. (2012) *The informal economy: Definitions, theories and policies.* Working Paper 1. Cambridge, MA: WIEGO.

Chok, S., Macbeth, J., & Warren, C. (2007). Tourism as a tool for poverty alleviation: A critical analysis of "pro-poor tourism" and implications for sustainability. *Current Issues in Tourism*, 10(2–3), 144–116. https://doi.org/10.2167/cit303.

Croes, R., and Vanegas, M. (2008). Cointegration and Causality between Tourism and Poverty Reduction. *Journal Of Travel Research*, 47(1), 94–103. http://dx.doi.org/10.1177/0047287507312429.

Dalit Welfare Organization (DWO) (2018). *Definition of Dalit in Nepal*, https://dwo.org.np/v1/dalit.php#:~:text=Dalit%20are%20the%20de%20facto,%2C%20language%2C%20culture%20and%20castes.

Espiner, S., Stewart, E. J., & Lama, L. T. (2017). Assessing the effectiveness of 'Appreciative Inquiry'(AI) in Nepali pro-poor tourism (PPT) development processes. *Tourism planning & development*, 14(3), 369–388.

Goodwin, H. (1999). *Sustainable tourism and poverty elimination study: A report to the department for international development.* Department for International Development.

Goodwin, H. (2008). Pro-poor tourism: A response. *Third World Quarterly*, 29(5), 869–871.

Goodwin, H. (2009). Contemporary policy debates: Reflections on 10 years of pro-poor tourism. *Journal of Policy Research in Tourism, Leisure and Events*, 1(1), 90–94.

Harrison, D. (2008). Pro-poor tourism: A critique. *Third World Quarterly*, 29(5), 851–868.

Hillman, W. (2022). Transitioning into Tourism: Becoming a Trekking Guide. *Journal of Tourism and Himalayan Adventures*, 4(1), 94–106.https://doi.org/10.3126/jtha.v4i1.46312.

Hillman, W. (2019a). "Risky business": The future for female trekking guides in Nepal. *Tourism and Hospitality Research*, 19(4), 397–407. https://www.jstor.org/stable/26789040.

Hillman, W. (2019b). Issues for women trekking in Nepal:'3 sisters adventure trekking' guides. *Journal of Sport & Tourism*, 23(1), 1–18. https://doi.org/10.1080/14775085.2019.1593226.

Hillman, W., & Radel, K. (2022a). The social, cultural, economic and political strategies extending women's territory by encroaching on patriarchal embeddedness in tourism in Nepal. *Journal of Sustainable Tourism*, 30(7), 1754–1775. https://doi.org/10.1080/09669582.2021.1894159.

Hillman, W., & Radel, K. (2022b). Transformations of Women in Tourism Work: A Case Study of Emancipation in Rural Nepal. *The Gaze: Journal of Tourism and Hospitality*, 13(1), 27–49. https://doi.org/10.3126/gaze.v13i1.42040.

Hummel, J., & van der Duim, R. (2012). Tourism and development at work: 15 years of tourism and poverty reduction within the SNV Netherlands Development Organisation. *Journal of Sustainable Tourism*, 20(3), 319–338. https://doi.org/10.1080/09669582.2012.663381.

Hummel, J. (2004). Pro-Poor Sustainable Tourism. In K. Luger, C. Baumgartner & K. Wöhler, S.Höivik, and B. Mayr (Eds.), . *Ferntourismus wohin? Der globale Tourismus erobert den Horizont* (123–146). Österr. StudienVerlag.

Jackson, M. A. (2010). *Empowering women of Nepal: An experience of empowerment in the land of the Himalaya* (Doctoral dissertation, Prescott College).

Kafle, J. (2011). *Poverty Alleviation through Sustainable Tourism Development in Nepal: Marketing Strategy Point of View.* Thesis Fall 2011 Business School International Business Management, Faculty: Business School Degree programme: International Business Management, Seinäjoki University of Applied Sciences, Finland.

Lemelin, R. H., Koster, R., & Youroukos, N. (2015). Tangible and intangible indicators of successful aboriginal tourism initiatives: A case study of two successful aboriginal tourism lodges in Northern Canada. *Tourism Management*, 47, 318–328. doi: http://dx.doi.org/10.1016/j.tourman. 2014.10.011.

Pandey, R. B. (2008). What should we do to practice Community Based Pro-poor Tourism: Lessons to be learned from TRPAP. In: P. Sharma (Ed.), *Vision and Mission of Village Tourism (32-38). Kathmandu: Village Tourism Promotion Forum.*

Nepal, S. K. (2022). The Quest for Sustainable Tourism in Nepal. *Current History*, 121 (834), 147–153. https://doi.org/10.1525/curh.2022.121.834.147.

Hall, D. R., & Richards, G. (Eds.). (2000). *Tourism and sustainable community development*. Routledge.

Roe, D., Goodwin, H., & Ashley, C. (2004). Pro-poor tourism: benefiting the poor. In T. V. Singh (Ed.), *New horizons in tourism: Strange experiences and stranger practices* (147–161). Cabi. https://doi.org/10.1079/9780851998633.0147.

Scheyvens, R. (2009). Pro-poor tourism: is there value beyond the rhetoric? *Tourism Recreation Research*, 34(2), 191–196. https://doi.org/10.1080/02508281.2009.11081590.

Scheyvens, R. (2007). Exploring the tourism-poverty nexus. *Current Issues in Tourism*, 10(2–3), 231–254.https://doi.org/10.2167/cit318.0.

Scheyvens, R. (1999). Ecotourism and the empowerment of local communities. *Tourism Management*, 20(2), 245–249. https://doi.org/10.1016/S0261-5177(98)00069-7.

Shah, K., & Gupta, V. (2000). *Tourism, the poor and other stakeholders: Experience in Asia*. Overseas Development Institute.

Saufi, A., O'Brien, D., & Wilkins, H. (2014). Inhibitors to host community participation in sustainable tourism development in developing countries. *Journal of Sustainable Tourism*, 22(5), 801–820. https://doi.org/10.1080/09669582.2013.861468.

Sinha, S. & Rishad, K. T. (2016). Tourism as a poverty alleviation tool in community building – A study on Arunachal Pradesh, *International Journal of Current Research*, 8(12), 43929–43934.

Seif, J. (2004). *Fair Trade in Tourism South Africa*. Presentation at ATLAS Conference, University of Pretoria, South Africa. November 2004.

Shah, K., & Gupta, V. (2000). *Tourism, the poor and other stakeholders: Experience in Asia*. London: Overseas Development Institute.

Shrestha, B., & Shrestha, A. (2021). Assessment of nature guide's perception towards biodiversity conservation in Chitwan national park using knowledge, attitude and practice (KAP) Approach. *Journal of Environmental Sciences*, 7, 61–73.

Shrestha, A. K. (1999). *Government Policies for Pro-Poor Growth: An Analysis of post-1990 Nepal*. Institute of Social Studies.

Spenceley, A., & Meyer, D. (2012). Tourism and poverty reduction: Theory and practice in less economically developed countries. *Journal of Sustainable Tourism*, 20(3), 297–317. https://doi.org/10.1080/09669582.2012.668909.

Tosun, C., & Jenkins, C. L. (1998). The evolution of tourism planning in third-world countries: A critique. *Progress in Tourism and Hospitality Research*, 4, 101–114. https://doi.org/10.1002/(SICI)1099-1603(199806)4:2<101::AID-PTH100>3.0.CO;2-Z.

Tucker, H., & Boonabaana, B. (2012). A critical analysis of tourism, gender and poverty reduction. *Journal of Sustainable Tourism*, 20(3), 437–455. https://doi.org/10.1080/09669582.2011.622769.

Upadhayaya, P. K., & Upreti, B. R. (2011). *Mountain tourism in Nepal: An overview on the sustainable inclusion of local communities.* Kathmandu: Mountain forum. https://lib.icimod.org/record/9484.

Zedong, M. (n.d.). "*Women hold up half the sky*".

Zeng, B., & Ryan, C. (2012). Assisting the poor in China through tourism development: A review of research. *Tourism Management*, 33, 239–248. https://doi.org/10.1016/j.tourman.2011.08.014.

Zhao, W., & Ritchie, J. B. (2007). Tourism and poverty alleviation: An integrative research framework. *Current issues in Tourism*, 10(2–3),119–143. https://doi.org/10.2167/cit296.0.

Part II

Impacts and Challenges of Taking a Pro-Poor Approach to Mountain Tourism

8 Socio-Cultural Integration in the Development of Pro-Poor Mountain Tourism

A Lesson from Eastern Turkey

Aysu Uğurlar and Mehmet Şeremet

Introduction

Much research has been done on pro-poor tourism since the early 2000s (Ashley et al., 2001; Çetin, 2012; Goodwin, 2008; Higgins-Desbiolles, 2008; Uğurlar, 2007). The increased interest in pro-poor tourism might be related to adopting development goals and their relation to sustainability. The UN Sustainable Development Goals (SDGs), in particular, have raised awareness in many communities worldwide of the importance of promoting equality, community development, and integration (Şeremet et al., 2021). Yet, while much research has focused on the role of pro-poor tourism in promoting community development, gender equity, and urban areas (Ashley & Roe, 2002; Higgins-Desbiolles, 2008; Karim et al., 2020; Li et al., 2015; Rogerson, 2006; Saayman & Giampiccali, 2015; Zeng & Ryan, 2012), there is little research on extending the pro-poor paradigm to the context of mountain communities – an extension that deserves attention for its potential to advance SDG 1 (end poverty) and SDG 10 (reduce inequalities) (Karim et al., 2020; Mutana & Mukwada, 2020). This study focuses on pro-poor tourism's role in integrating a nomad community into society and safeguarding the sustainability of mountainous areas.

This study selected the eastern part of Turkey, a mountainous region populated by Kurdish communities, as a case study area. While western Turkey, with its "sea-sun-and-sand" potential for Western tourists, has long been a popular tourist destination, the eastern part has recently also received attention due to its undiscovered unique cultural and natural assets (Alvarez, 2010; Cihangir & Şeremet 2020; Efe et al., 2022). Given that Lake Van has been host to many civilisations in the past (Urartian, Assur, Seljuk, Armenian, Ottoman), the region shows a mix of these heritages (Şeremet et al. 2022). However, due to deficient border security and terrorist attacks, the region has been less safe for national and international tourists. With the end of terrorism, this region has entered a new development stage with governmental planning and a policy of tourism development (Cihangir & Şeremet, 2020), becoming a cultural tourism destination as part of the Van Lake Basin and Urartian cultural and historical corridor (Şeremet et al., 2022). Such

DOI: 10.4324/9781003392583-10

tourism policy and planning should, however, be critically evaluated through a pro-poor tourism framework, which might offer some insights into the challenges and opportunities for the future development of sustainable tourism in the region, which deserves both social and economic development to reduce inequalities as well as the cultural integration of the local communities with each other.

The pro-poor tourism framework offers insight into how deprived communities gain economic and social power. This framework was constructed by scholars who focus on comprehensiveness, sustainability, community involvement and development, capacity development, the sharing economy, and cultural and natural awareness (Ashley & Roe, 2002; Das & Chatterjee, 2015; Goodwin, 2008; Hall, 2007; Harrison, 2008; Rogerson, 2006). Therefore, many relevant concepts and theories were used to enhance the local communities' wellness and livelihood. Nevertheless, social integration in mountainous areas has received little attention as it is significant to rural areas' development while avoiding spoiling local culture and natural assets. This study, therefore, aims to discover the experiences of adapted communities in tourism, which can be considered pro-poor tourism in the context of mountainous environments. The following are the main research questions that this study will answer:

- What is the role of pro-poor tourism approaches in integrating immigrant societies?
- What are the challenges faced by immigrant societies in the implementation of pro-poor tourism?
- What advantages and opportunities does it offer for the individual development of immigrants during the realisation stages of pro-poor tourism?
- How does pro-poor tourism contribute to communities' adaptation to wider society?

To address these questions, we have utilised a qualitative research design to explore the ethnographic experiences of local people seeking to embrace rural communities' socio-cultural environment. This study focuses on Kyrgyz communities residing in eastern Turkey, employing a qualitative approach.

Literature Review

Poverty and pro-poor tourism

In the 1970s, the concept of sustainability emerged from environmentalism, focusing on the negative effects of human life on the natural balance of the world (Emrealp, 2005). Sustainable development, based on economic development and the protection of social and environmental values, became a goal at local, national, regional, and global levels (Liu, 2003; Uğurlar, 2007; Bâc,

2014). International organisations, such as the International Labor Organization and the World Bank, discussed the relationship between poverty and sustainability. The World Tourism Organization defined sustainable tourism as "leading the management of all resources to meet economic, social, and aesthetic needs while preserving cultural integrity, fundamental ecological processes, biodiversity, and life support systems." (WTO, 1998, as cited in Ashley et al. 2001; Harrison 2008). Sustainable tourism development has evolved since the early 1980s, with eco-tourism, alternative tourism, small-scale, responsible tourism, soft tourism, and community-based tourism being promoted as pathways to sustainable development (Ashley & Roe, 2002; Kantsperger et al., 2019; Liu, 2003). The adoption of Agenda 21 at the Rio Summit in 1992 marked a turning point, with the World Travel and Tourism Council adopting the principle of meeting the needs of today's tourists and host regions while preserving and enhancing future opportunities (Harrison, 2008; Uğurlar, 2007). The Millennium Development Goals of 2000 and the UN's Agenda 2030 aspire to eradicate poverty in all its forms as a component of long-term development (Halisçelik, 2015; Şeremet et al., 2021).

Paradigm shifts in pro-poor tourism

In light of the evolving paradigms surrounding poverty in literature, poverty reduction is a primary concern in development policy, representing a multidimensional concept (Kakwani & Pernia, 2000). Following the discourse on sustainability and poverty that emerged in the 1990s, the notion of "pro-poor growth" emerged in discussions during the early 2000s (Kakwani & Pernia, 2000; Uğurlar, 2007). In broad terms, "pro-poor growth" denotes growth anticipated to assist individuals in poverty (Ravallion, 2004), enabling their active participation in economic activities and deriving significant benefits from them (Kakwani & Pernia, 2000). Poverty alleviation focuses on improving human well-being, particularly that of impoverished individuals (Kakwani & Pernia, 2000). In the past, sustainable tourism development, now recognised as a tool for poverty alleviation, was primarily focused on macroeconomic growth and maximising profits for the private sector (Torres & Momsen, 2004). Since the 1960s, tourism has been an integral part of economic development strategies, especially in developing countries (Torres & Momsen, 2004). The tourism sector's potential to attract international investment, generate foreign exchange earnings, create new employment opportunities, and catalyse macroeconomic growth has served as an incentive (Torres & Momsen, 2004). In other words, the high multiplier effect of tourism can make tourism the economic driver of a nation (Fang, 2020). Torres and Momsen (2004) state that tourism's potential for poverty reduction has been largely overlooked. According to Schilcher (2007), while poverty reduction strategies are subject to ideological interpretations within (capitalist) extremes such as neoliberalism and protectionism, Rogerson (2006) argues that the tourism sector, driven by purely commercial interests, has many advantages over other sectors in the fight against poverty, despite its limited capacity to serve people experiencing poverty. Tourism is a powerful tool for poverty reduction and can bring socio-economic benefits to various places (Tolkach et al., 2012). Therefore, with the rise of neoliberalism, which prioritises outward-oriented economic development techniques, Southern countries are increasingly turning to tourism as an economic development strategy (Torres & Momsen, 2004). In contrast, some scholars (Britton, 1981; Brohman, 1996; Crick, 1989 cited in Torres & Momsen, 2004: 295) have conceptualised tourism in the Third World as a form of exploitative "neocolonialism" or "leisure imperialism".

The predominant argument in the literature concerning poverty and tourism studies is that conventional investment strategies fall short of effectively addressing the widespread issue of poverty (Ashley & Haysom, 2006; Ashley & Roe, 2002). In the pursuit of poverty reduction, it becomes imperative to achieve growth that encompasses the impoverished and furnishes them with tangible benefits (Ashley & Haysom, 2006; Ashley & Roe, 2002). While poverty reduction may not be the central focus of the tourism agenda, research examining the relationship between tourism and poverty suggests that tourism holds significant potential in combating poverty (Ashley et al., 2001; Zhou et al., 2016). Fang (2020) contends that tourism can effectively diminish community poverty, enhance residents' well-being, and eradicate poverty. However, Liu (2003) suggests that global experience reveals a limited realisation of this tourism potential. In contrast, Harrison (2008) points out the dearth of empirical evidence on the impact of tourism in reducing poverty. Fang (2020) emphasises the effectiveness of tourism in addressing poverty but stresses the need for long-term strategies and the widespread distribution of its benefits across society. The inception of the "pro-poor tourism" movement, traced back to the late 1990s, contextualises this development (Chok et al., 2007). In the context of the Rio Conference on Environment and Development in 1992, the Department for International Development (DFID) prepared a report on "Tourism and Poverty Elimination" in preparation for the UN Commission on Sustainable Development session in New York in 1999. The report seeks to answer how tourism can impact poverty reduction, assuming that tourism to developing countries is important and growing and that this trend will continue. The report addresses how the economic benefits of international tourism in developing countries can be maximised through employment and micro-enterprise opportunities. It also includes a discussion of tourism's negative economic, social, and environmental impacts to maximise the benefits. The report argues for increased local participation in planning and governance and for ensuring that poorer members of society secure market access and property rights, but it also acknowledges the neoliberal paradigm. From this perspective, the notion of pro-poor tourism, introduced in the report, is characterised as tourism that yields net benefits exceeding costs for impoverished individuals (Ashley et al., 2001). These benefits may be economic and social, environmental, or cultural (Ashley et al., 2001). Ashley et al. (2001)

note that the poor-oriented tourism approach focuses on strategies that increase the benefits provided to people experiencing poverty rather than defining net positive benefits. In other words, pro-poor tourism is a potentially comprehensive approach that provides economic gains for poor people, non-economic livelihoods, and less tangible benefits (Peeters, 2009; Truong, 2018). The tourism sector approach includes many stakeholders at all levels, from micro to macro. Stakeholders include the government, the private sector, civil society, and people with low incomes, acting as producers and decision-makers (Ashley et al., 2001; Tolkach et al., 2012). The pro-poor tourism approach is a development methodology that aims to use tourism as a tool to reduce poverty (Gascon, 2015), as well as tourism development and management (Fang, 2020). The main aim of the pro-poor tourism strategies is to unlock opportunities for low-income people rather than expand the sector (Ashley et al., 2001; Knight, 2018; Zhou et al., 2016). Pro-poor tourism needs to connect with the mainstream (DFID, 1999, as cited in Goodwin, 2008). Pro-poor tourism cannot succeed without successfully developing the entire tourism destination (Ashley et al., 2001; DFID, 1999, as cited in Goodwin, 2008). According to Goodwin (2008), the pro-poor tourism approach, which is based on interaction with the mainstream tourism sector, recognises that in cases where tourism is sufficiently scaled in poverty reduction, it can be effective when households can contribute adequately to their income. This differentiates the pro-poor tourism approach from the eco-tourism (conservation and wildlife-oriented) and community-based tourism (community benefit) approaches, which focus less on poverty reduction (Goodwin, 2008). In essence, pro-poor tourism intersects with both eco-tourism and community-based tourism (CBT) to some extent (Ashley et al., 2001). However, the pro-poor tourism approach is significantly aligned with fair, just and equitable tourism approaches. However, the main difference is that pro-poor tourism initiatives primarily focus on initiatives in the Southern destinations in South Africa (Ashley et al., 2001). In short, pro-poor tourism, touted as an "alternative" approach to mainstream tourism development models, is geared towards providing net benefits for the poor (Chok et al., 2007) and focuses more on countries of the South, not on mainstream destinations in the North. Poverty is the core focus, rather than one element of (mainly environmental) sustainability (Ashley et al., 2001).

Pro-poor tourism approaches and strategies can be applied in urban, rural, and coastal regions (Ashley & Roe, 2002). Pro-poor tourism is not just about community tourism and small, medium, and micro enterprises (SMEs), nor is it just about tourism in poor areas, nor is it a niche product sold as "interaction with local people", although such initiatives can contribute to pro-poor tourism (Ashley & Haysom, 2006). Pro-poor tourism focuses on creating opportunities for people experiencing poverty in the tourism sector and enabling them to benefit from tourism in various ways, such as employment, income generation, skills development, and overall economic empowerment. Inclusive and equitable tourism practices are needed to address poverty and social inclusion on a broader scale (Ashley & Haysom, 2006; Ashley & Roe, 2002).

Pro-poor tourism strategies fall into three main groups (Ashley & Roe, 2002; Zhou et al., 2016). The first is the expansion of economic benefits for the poor. This strategy involves increasing local employment and local enterprise opportunities by providing suitable training and generating collective income that spreads to the broader community. The second is addressing livelihood impacts. This strategy includes capacity building and training, reducing environmental impacts from tourism, addressing the competitive use of natural resources, improving social and cultural impacts (such as preventing social degradation), and increasing access to infrastructure and services. The third is the development of pro-poor policies, processes and partnerships that overlap with the above two strategies. This strategy includes creating a more supportive policy and planning framework that removes barriers to the participation of people with low incomes, increasing their participation in decision-making processes, encouraging pro-poor partnerships with the private sector in developing new tourism products, and increasing the flow of information communication (Ashley & Roe, 2002; Zhou et al., 2016).

Despite shifts in the meaning of pro-poor terms, the pro-poor tourism approach remains a valuable framework for prioritising the gains that help the most disadvantaged people in any given society (Spenceley, 2022). However, emphasising the results of a few pro-poor tourism-related researches show that the pro-poor impact of pro-poor tourism projects has been limited due to a lack of participation, resulting in an uneven distribution of the benefits of tourism in local communities (Lo et al., 2019). Trau (2012) examines local communities' initiatives to reduce poverty in the context of pro-poor tourism, which is seen as a globalised tourism-based approach with a critical perspective in a study. According to the findings of this study, the collective ownership Lelema communities in the Land of Chef Roi Mata were able to translate the globalised pro-poor tourism development discourse into the local culture through Roi Mata Cultural Tours, exemplifying so-called "glocalisation." As a characteristic example of its theory, it can simultaneously reflect both global and local values of the local enterprise (Trau, 2012). The findings of Lo et al.'s (2019) study reveal that pro-poor tourism provides tangible economic benefits for local communities but can also create an essential gap in terms of income and capital equipment between participating and non-participating households because, effective participation is crucial and necessary to access the benefits from pro-poor tourism.

Furthermore, Solimans' (2015) study in the protected area context of Fayoum in Egypt asserted such problems as the lack of infrastructure and superstructure and lack of involvement of local people in the development process, which are principal barriers to pro-poor tourism development. The study of Torabi et al. (2019) focuses on the benefits of pro-poor tourism to Iran's intangible and tangible assets. In this context, they identify the issues and challenges of pro-poor tourism, focusing on a tourism development program (TEPR) implemented in three selected villages of Shahrud County, Iran. The results of this study suggest that the financial and non-financial support under the tourism, employment, and poverty reduction (TEPR) program is provided more to the private sector and dominant social groups than to the poor. Although many mountain communities are faced with the reality of poverty in developing countries (Mutana & Mukwada, 2020) and benefit from its adventure potential for tourism (Apollo & Wengel, 2022), surprisingly, there are almost no studies on pro-poor tourism in mountainous areas.

In contrast, as current research in the context of pro-poor tourism research in mountain areas, Karim's et al. (2020) study focuses on some of the critical pro-poor tourism initiatives adopted for community-based development in the Hunza Valley of northern Pakistan. Based on the results, the research argues that pro-poor tourism must be integrated into a broader community-based development strategy to be truly effective in disadvantaged communities. Mutana and Mukwada (2020) conducted surveys and in-depth interviews with tourism operators and local government officials to assess the contribution of route tourism to poverty reduction along the Maluti Route in South Africa's Drakensberg Mountain Region. According to the study's findings, commercial organisations believe that they have contributed significantly to poverty reduction in the region, while local communities acknowledge only limited benefits from tourism. The main result of this work was the need to communicate among stakeholders about current efforts and options to reduce poverty through tourism (Mutana & Mukwada, 2020). Similarly, the primary finding of the research conducted by Wen et al. (2021) demonstrate that reducing poverty through strategic tourism development that neglects effective collaboration among the three key stakeholders—local perspective, tourism enterprises, and community residents—will not yield substantial benefits for the impoverished.

The pro-poor tourism literature has been instrumental in proposing solutions for impoverished communities. However, critical studies on pro-poor tourism have emerged. Chok et al. (2007) contend that in fragmented and environmentally vulnerable communities, dependence on tourism as a tool for poverty reduction is unlikely to yield substantial long-term benefits for those already marginalised. According to Harrison (2008), pro-poor tourism provides no unique method and is neither anti-capitalist nor opposed to mainstream tourism. In this context, Harrison (2008) asserts that pro-poor tourism should be reintegrated into mainstream tourism and development research to underscore the exploration of its actual and potential roles in poverty reduction and promoting "development". Peeters (2009), however, argues that pro-poor tourism does not significantly contribute to poverty alleviation and may even impede efforts to address climate change. Gascón (2015), on the other hand, criticises the definition of pro-poor tourism as solely "net income-based," contending that it inadequately addresses rural poverty issues. This highlights the notion that poverty cannot solely be confined to economic and social benefits, necessitating consideration of other social factors such as isolation, cultural adaptation, and intercultural communication. This study primarily focuses on a community striving for recognition and inclusion in poverty reduction efforts within a mountainous area, specifically the Ulupamir Kyrgyz Village of Van, Turkey.

Findings

We found several issues to be discussed under the framework of pro-poor tourism, which allows us to analyse the qualitative data.

Tourism as part of experience in comprehensiveness

Although mountain poverty may not inherently differ from lowland poverty, Gerlitz et al. (2012, as cited in Mutana & Mukwada, 2020) strongly argue that poverty experienced in mountain communities is more severe due to the heightened exposure to both environmental and non-environmental shocks. In this case study, the local inhabitants expressed a sense of detachment from the land and the region upon their arrival (see Figure 8.1). Despite possessing some cultural and social capital, their initial adaptation to the region was challenging, particularly given that their primary skills revolved around

Figure 8.1 A view of Ulupamir Kyrgyz Village, transitioning from nomadic to sedentary life in Van, Turkey

husbandry and animal care. Moreover, the mountainous terrain presented harsh winter conditions and a scarcity of flat surfaces, crucial for their socio-cultural activities, primarily centred around horsemanship on expansive lands (see Figure 8.1). Consequently, most festivals encompass sports, social gatherings, and cultural events intricately linked to horses, symbolising "freedom" in their lifestyle. However, their cultural practices were often constrained by the environmental conditions, impacting them during times of restriction. Therefore, the emergence of tourism as a viable economic opportunity allows them to develop their cultural capital and organise festivals more effectively.

Experience in sustainability

Increasingly common spatial limitations, lack of institutional support for tourism activities, and the necessity of seeking living spaces outside the village among the third generation are emerging trends. These factors may pose threats to the preservation of local languages and traditions. While the government has offered some opportunities for them to adapt to local culture and economic life, tourism has spontaneously emerged as an economic asset to support the sustainability of local cultural life as well as contribute to the principles of sustainability in the context of mountain life. The prominent matter is probably related to the sustainability of their intangible cultural assets such as festivals, wedding ceremonies, sportive competitions, and local products (gastronomy, production of carpet, music) (Figure 8.2). However, since transferring these assets is difficult in a mountain-based environment, this also created some challenges for them and increased their isolation. For example, the fermented milk of mares is a special drink for them, which is called "Kmmz (Kummis)," and eating horse meat is a cultural thing. However, the surrounding Kurdish community primarily consumed lamb meat and collected milk to produce "herby cheese," a staple for surviving the harsh winter conditions. Crafting herby cheese holds significant socio-ecological importance for local women in the region (Cihangir et al., 2020). Initially, their cultural practices were constrained by unsuitable environmental conditions and social capital. Thus, tourism has provided them another avenue to uphold and share their cultural traditions with the local communities.

Involvement, participation and capacity development

Upon reviewing the policy document, it became apparent that the Regional Development Agency (DAKA) assigned relatively little importance to the tourism development plan for the Kyrgyz village, resulting in a lack of direct support. Nonetheless, a local entrepreneur took the initiative to establish an area where both local and international tourists could experience the local culture, named "Atayurt" (motherland) (see Figure 8.3). Although initially situated in the village, it later had to be relocated, facilitating the adaptation of local culture and its dissemination to surrounding communities. Initially,

Figure 8.2 A view of women's handicrafts (carpentry) in Ulupamir Kyrgyz Village

Figure 8.3 A view of the tourism enterprise which is called "Atayurt" (Motherland)

the human capital of the local population was predominantly utilised in government jobs and the service sector, rendering their technical skills less relevant for the sustainability of the local community. However, given the significance of critical geopolitics in shaping local life (Cihangir et al., 2022), Turkey's proximity to Russia and central Asian countries proved advantageous for the local community. The region's proximity to the Malazgirt battlefield and the Seljuk cemetery in Ahlat province led to the initiation of an annual festival aimed at promoting Turkish cultural practices. The local population's technical skills in producing traditional clothing and tents were crucial. As a result, the rising interest in historical costumes, spurred by a growing number of television series, gave rise to a new industry for costume production within the village.

The constrained capacity of the regional economy to meet the demand for cinema sector design represents a promising development, potentially acting as a catalyst for regional economic growth. However, despite latent or robust non-institutional local engagement, neither the local populace nor the most disadvantaged segments of society are granted the authority to engage actively in activities associated with tourism. However, educational opportunities enable women to receive education and pursue career advancement outside the village. Conversely, those who remained in the village had fewer options to support themselves, often limited to joining the army or securing low-paying jobs in carpet studios. Therefore, tourism could provide employment opportunities while allowing them to engage in and preserve their cultural heritage inherited from their families.

Cultural and natural awareness

As a migrant society from Central Asia, the local community has historically strived to strike a balance between Kurdish and Turkish influences, making geopolitics an integral aspect of their daily lives. Consequently, as a Turkish village amidst a predominantly Kurdish population, they face challenges related to integration, communication, and preserving cultural traditions. However, they have recently embraced tourism and cultural activities to garner attention from broader communities. With the TV series and the encouraging effect of the "Atayurt" activity field, the village has started to be visited by many people. However, the village does not appear to be transitioning into a tourism destination due to a lack of support and training for the local population. Several social-structural issues are relevant to the tourism development process, particularly conservatism. The senior Kyrgyz members, who predominantly lead the community, guide the younger generations.

Consequently, apart from routine sporting and cultural activities, the locals are hesitant to endorse the establishment of tourist facilities within the village, resulting in limited host-tourist interaction. Despite the presence of a hydroenergy centre that may threaten the surrounding natural environment, the villagers do not voice complaints about its existence. This may indicate that they view the centre as representing state authority, especially considering that many villagers hold governmental positions in the province and city centre. Consequently, they prefer not to engage in conflicts with the government.

Discussion and Conclusion

This chapter examines the potential for pro-poor tourism within the migrant Kyrgyz community residing in a mountainous region of Turkey. Initially, the community received governmental support through social policies such as housing and employment opportunities. However, as subsequent generations seek improved living conditions and opportunities in larger cities, education emerges as a critical social policy support for the younger generation. This shift is evident in the significant migration of young people from the village to urban areas. Additionally, geopolitical changes in the region have presented opportunities for the local community to showcase their cultural and intangible assets.

As a result, community members have gained avenues for both preserving their cultural heritage and enhancing their income. On the plus side, the limited tourism activities enhance cultural interaction between the migrant community and the residents, facilitated by a single tourist facility. Entrepreneurs within the village are initiating tourism ventures, thereby creating job opportunities for the locals. However, while the community seeks opportunities for social integration, economic benefits may not be their primary focus. As many scholars have argued that the perception of poverty is subjective, Karim et al. (2020) also contend that tourism can only contribute to the well-being of impoverished individuals. Thus, this study demonstrates that the cultural capital of the local population is being widely deployed and shared through initial tourism activities. This presents further potential for the future of tourism in the community, including horseback riding, local festivals, and sports tournaments. These activities have the potential to serve as tourist attractions through which the local people can interact with visitors and other members of the community.

On the less positive side, the senior community has not welcomed the tourism activities within the village. Given that the young generation takes advantage of education opportunities, they tend to leave the village to live in big cities, pursuing better opportunities and social life. This might be a challenge for the village's human resources for growing as a tourism village in the future as such learned skills as horse riding, fermenting kumis, playing sports games, and cooking the local traditional food can only be sustained by the local people who grew up in this culture. Given that deprivation has a variety of dimensions (Bakker & Messerli, 2017), socio-cultural isolation still exists in the region. Without intercultural communication, being part of an unintegrated community forces them to be a closed society.

Regarding gender inequality, women in the village appear to be either unpaid labour or employed in low-wage positions. Although young girls are permitted to pursue education, women in the village continue to bear the burden of household chores, with only a small group engaged in rug weaving for income. Consequently, women in the village feel poverty more acutely than men.

In conclusion, the case study of mountainous communities in eastern Turkey highlights the potential of pro-poor tourism to facilitate social interaction and communication, provided that adequate policy and planning support is extended to local communities. This study underscores that poverty is not solely an economic issue but also encompasses cultural integration, intercultural communication, and social integration. However, despite the positive impact of pro-poor tourism on reducing poverty and creating job opportunities (Scheyvens & Russell, 2012), this study indicates that the primary benefit for poor communities stems from the recognition they receive through tourism. This could also be attributed to the significant connection between geopolitics and tourism (Cihangir et al., 2022). As Torres and Momsen (2004) focus on the drivers of the development, fostering of socio-ecological skills and its turning into tourism potential needs to be supported by the local agencies which might offer the vulnerable groups project support as well as financial support to small tourism entrepreneurs.

Management Implications

This chapter also draws some broader lessons from the Turkish experience as well as managerial implications. Given that socio-ecological skills and knowledge were significant characteristics of the local communities, the area managers should adopt collaboration strategies that bring managing policies and incentives, adapting the local communities' potential cultural practices and socio-cultural background.

Overall, policymakers must implement policy and planning interventions to support community-driven tourism development in the region and alleviate deprivation. This can be achieved through the following actions:

- To safeguard the mountaineering environment, cultural assets and natural resources should be integrated, for example via a "geopark project."
- NGO support should be provided to the local populace to facilitate the development of sustainable mountaineering tourism opportunities, guided by a pro-poor tourism approach.
- Poverty should be addressed by acknowledging its multidimensional nature. In rural communities, the rules and regulations should be adapted to the local circumstances to establish a mechanism facilitating access to opportunities.

References

Alvarez, M. D. (2010). Marketing of Turkey as a tourism destination. *Anatolia*, 21(1), 123–138. https://doi.org/10.1080/13032917.2010.9687094.

Apollo, M., & Wengel, Y. (2022). *Mountaineering tourism: A critical perspective*. Routledge. https://doi.org/10.4324/9781003095323.

Ashley, C., & Haysom, G. (2006). From philanthropy to a different way of doing business: Strategies and challenges in integrating pro-poor approaches into tourism business. *Development Southern Africa*, 23(2), 265–280. https://doi.org/10.1080/03768350600707553.

Ashley, C., & Roe, D. (2002). Making tourism work for the poor: Strategies and challenges in Southern Africa. *Development Southern Africa*, 19(1), 61–82. https://doi.org/10.1080/03768350220123855.

Ashley, C., Roe, D., & Goodwin, H. (2001). *Pro-poor tourism strategies: Making tourism work for the poor – A review of experience*. Overseas Development Institute.

Bâc, D. P. (2014). The emergence of sustainable tourism – A literature review. *QUAESTUS Multidisciplinary Research Journal*, 4, 131–140.

Bakker, M., & Messerli, H. R. (2017). Inclusive growth versus pro-poor growth: Implications for tourism development. *Tourism and Hospitality Research*, 17(4), 384–391. https://doi.org/10.1177/1467358416638.

Das, M., & Chatterjee, B. (2015). Eco-tourism: A panacea or a predicament? *Tourism Management Perspectives*, 14, 3–16. https://doi.org/10.1016/j.tmp.2015.01.002.

Chok, S., Macbeth, J., & Warren, C. (2007). Tourism as a tool for poverty alleviation: A critical analysis of 'pro-poor tourism' and implications for sustainability. *Current Issues in Tourism*, 10(2–3),144–165. https://doi.org/10.2167/cit303.

Cihangir, E., Demirhan, Ö., & Şeremet, M. (2020). Kırsal turizm potansiyelinin ortaya çıkarılmasında kırsalda kadının durumuna bir bakış: Van Otlu Peyniri üretimi örneği. *International Journal of Geography and Geography Education*, (42), 199–222. https://doi.org/10.32003/igge.692207.

Cihangir, E., & Şeremet, M. (2020). Van'ın Edremit ilçesinde turizmin gelişimi: Rezilyans ve değişim. *Türk Coğrafya Dergisi*, (74), 47–59. https://doi.org/10.17211/tcd.657017.

Cihangir, E., Şeremet, M., & Cihangir-Çamur, K. (2022). Turkey at the crossroads: A study of geopolitics and tourism re-alignment. *Geography*, 107(3), 145–152. https://doi.org/10.1080/00167487.2022.2114165.

Çetin, Z. Ö. (2012). Yoksulluğu azaltmada yeni bir yaklaşim: Yoksul-yanlı turizm. *Pamukkale Üniversitesi Sosyal Bilimler Enstitüsü Dergisi*, (11), 97–108.

Efe, B., Gözet, E., Özgür, E., Lupo, A. R., & Deniz, A. (2022). Spatiotemporal variation of tourism climate index for Türkiye during 1981–2020. *Climate*, 10(10), 151. https://doi.org/10.3390/cli10100151.

Emrealp, S. (2005). *Yerel Gündem 21 uygulamalarına yönelik kolaylaştırıcı bilgiler el kitabı*. Iula-Emme Yayını.

Fang, WT. (2020). Pro-poor tourism. In *Tourism in emerging economies*. Springer. https://doi.org/10.1007/978-981-15-2463-9_7.

Gascón, J. (2015). Pro-poor tourism as a strategy to fight rural poverty: A critique. *Journal of Agrarian Change*, 15(4), 499–518. https://doi.org/10.1111/joac.12087.

Goodwin, H. (2008). Tourism, local economic development, and poverty reduction. *Applied Research in Economic Development*, 5(3), 55–64.

Halisçelik, E. (2015). *Binyıl kalkınma hedefleri ve yoksul odaklı büyüme*. Gazi Kitabevi,

Hall, C. M. (2007). *Pro-poor tourism: Who benefits? Perspectives on tourism and poverty reduction* (Vol. 3). Channel View Publications.

Harrison, D. (2008). Pro-poor tourism: A critique. *Third World Quarterly*, 29(5), 851–868. https://doi.org/10.1080/01436590802105983.

Higgins-Desbiolles, F. (2008). Justice tourism and alternative globalisation. *Journal of Sustainable Tourism*, 16(3), 345–364. https://doi.org/10.1080/09669580802154132.

Kakwani, N., & Pernia, E. M. (2000). What is pro-poor growth? *Asian Development Review*, 18(1), 1–16.

Kantsperger, M., Thees, H., & Eckert, C. (2019). Local participation in tourism development-roles of non-tourism related residents of the Alpine destination bad Reichenhall. *Sustainability*, 11(24), 6947. https://doi.org/10.3390/su11246947.

Karim, R., Muhammad, F., Serafino, L., Latip, N. A., & Marzuki, A. (2020). Integration of pro-poor tourism activities in community-based development initiatives: A case study in high mountain areas of Pakistan. *International Journal of Scientific and Technology Research*, 9(4), 848–852.

Knight, D. W. (2018). An institutional analysis of local strategies for enhancing pro-poor tourism outcomes in Cuzco, Peru. *Journal of Sustainable Tourism*, 26(4), 631–648. https://doi.org/10.1080/09669582.2017.1377720.

Li, H. Q., Hou, L. C., & Yang, S. W. (2015). Review of overseas research on pro-poor tourism. *Human Geography*, 30 (1), 26–32. https://doi.org/10.13959/j.issn.1003-2398.2015.01.005.

Liu, Z. (2003). Sustainable tourism development: A critique. *Journal of Sustainable Tourism*, 11(6), 459–475. https://doi.org/10.1080/09669580308667216.

Lo, K., Li, J., Wang, M., Li, C., Li, S., & Li, Y. (2019). A comparative analysis of participating and non-participating households in pro-poor tourism in Southern Shaanxi, China. *Tourism Planning & Development*, 16(3), 318–333. https://doi.org/10.1080/21568316.2018.1490340.

Manwa, H., & Manwa, F. (2014). Poverty alleviation through pro-poor tourism: The role of Botswana forest reserves. *Sustainability*, 6(9), 5697–5713. https://doi.org/10.3390/su6095697.

Mutana, S., & Mukwada, G. (2020). Can mountain route tourism work for the poor? Exploring worldviews from Maluti Route in the Drakensberg Region of South Africa. *Tourism and Hospitality Research*, 20 (1), 18–30. https://doi.org/10.1177/1467358418791312.

Peeters, P. (2009). Pro-poor tourism, climate change and sustainable development. *Tourism Recreation Research*, 34(2), 203–205. https://doi.org/10.1080/02508281.2009.11081593.

Ravallion, Martin, Pro-Poor Growth: A Primer (March 17, 2004). Available at SSRN: https://ssrn.com/abstract=610283.

Rogerson, C. M. (2006). Pro-poor local economic development in South Africa: The role of pro-poor tourism. *Local Environment*, 11(1), 37–60. https://doi.org/10.1080/13549830500396149.

Saayman, M., & Giampiccali, A. (2015). Community-based tourism and pro-poor tourism: dissimilar positioning in relation to community development. *Journal for New Generation Sciences*, 13(3), 163–181.

Scheyvens, R., & Russell, M. (2012). Tourism and poverty alleviation in Fiji: Comparing the impacts of small-and large-scale tourism enterprises. *Journal of Sustainable Tourism*, 20(3), 417–436. https://doi.org/10.1080/09669582.2011.629049.

Schilcher, D. (2007). Growth versus equity: The continuum of pro-poor tourism and neoliberal governance. *Current Issues in Tourism*, 10(2–3),166–193. https://doi.org/10.2167/cit304.0.

Şeremet, M., Haigh, M., & Cihangir, E. (2021). Fostering constructive thinking about the 'wicked problems' of team-work and decision-making in tourism and geography.

Journal of Geography in Higher Education, 45(4), 517–537. https://doi.org/10.1080/03098265.2020.1869924.

Şeremet, M., Cihangir, E., & Karaduman, E. (2022). (Non) branding cities and (De) institutionalisation perspectives: A case study of Van, Turkey. *Coğrafya Dergisi*, 45, 111–124. https://doi.org/10.26650/JGEOG2022-1100847.

Soliman, M. S. (2015). Pro-poor tourism in protected areas–opportunities and challenges: The case of Fayoum, Egypt. *Anatolia*, 26(1), 61–72. https://doi.org/10.1080/13032917.2014.906353.

Spenceley, A. (2022). Pro-poor tourism's evolution and implications arising from the COVID-19 pandemic. *Tourism Planning & Development*, 19(1), 13–25. https://doi.org/10.1080/21568316.2021.2021470.

Tolkach, D., Pearlman, M., & King, B. (2012). Key implementation factors in pro-poor tourism. *Tourism Recreation Research*, 37(1), 3–13. https://doi.org/10.1080/02508281.2012.11081683.

Torabi, Z. A., Rezvani, M. R., & Badri, S. A. (2019). Pro-poor tourism in Iran: The case of three selected villages in Shahrud. *Anatolia*, 30(3), 368–378. https://doi.org/10.1080/13032917.2019.1595689.

Torres, R., & Momsen, J. H. (2004). Challenges and potential for linking tourism and agriculture to achieve pro-poor tourism objectives. *Progress in Development Studies*, 4(4), 294–318. https://doi.org/10.1191/1464993404ps09.

Trau, A. M. (2012). Beyond pro-poor tourism:(re) interpreting tourism-based approaches to poverty alleviation in Vanuatu. *Tourism Planning & Development*, 9(2), 149–164. https://doi.org/10.1080/21568316.2011.630750.

Truong, V. D. (2018). Tourism, poverty alleviation, and the informal economy: The street vendors of Hanoi, Vietnam. *Tourism Recreation Research*, 43(1), 52–67. https://doi.org/10.1080/02508281.2017.1370568.

Uğurlar, A (2007). Sürdürülebilir turizm ve yoksulluk, In O. K. Armağan (Ed.), *Kent ve bölge üzerine çalışmalar* (455–468). Gazi Üniversitesi İletişim Fakültesi Basım Evi, Ankara.

Wen, S., Cai, X., & Li, J. (2021). Pro-poor tourism and local practices: An empirical study of an autonomous county in China. *Sage Open*, 11(2). https://doi.org/10.1177/21582440211022740.

Zeng, B., & Ryan, C. (2012). Assisting the poor in China through tourism development: A review of research. *Tourism Management*, 33(2), 239–248. https://doi.org/10.1016/j.tourman.2011.08.014.

Zhou, R., Zhong, L. S., Zhang, S. R., & Ma, X. Y. (2016, August). *Pro-poor tourism as approach to alleviation poverty: Experiences and critical issues* [Paper presentation] 3rd International Conference on Applied Social Science Research (ICASSR 2015) Beijing, China.

9 Mountain Tourism and Socio-Economic Outcomes in Mountain Communities in Nigeria

Prospects and Challenges

Kennedy Eborka, Ganiyu Lawal, Ibrahim Bidemi Abdullateef, Oluwayomi Adeleke and Temitope Owolabi

Introduction

The growing relevance of mountain tourism as a potential force for the economic transformation of mountain communities and national development continues to attract exciting academic discourses. Mountains cover about 24% of the world's total land area and constitute a popular object of tourist attraction (Richins et al., 2016). Studies indicate that mountain tourism accounts for about 15% to 20% of annual global tourism and currently represents the fastest-growing sector in the world, and is expected to generate up to 5,980 billion USD in revenue by 2031 (Birendra et al., 2021; Rasulovich, 2021; Taher et al., 2015). However, Africa's tourism and hospitality sector is underutilised, with about $203.7 billion in unrealised potential (Frechtling & Hara, 2016).

Despite the absence of a universally embraced definition, some scholars define the phenomenon by describing the activities people engage in to constitute mountain tourism. For instance, individuals or groups' local and international travels to mountain communities to realise some personal goal or self-achievement are described as mountain tourism (Ritpanitchajchaval et al., 2023). Taher et al. (2015) categorised mountain tourism into several activities, including mountain hiking, mountain trekking, and ice and rock hiking in mountain regions. However, the authors described these activities as mountain adventure rather than mountain tourism. Other scholars, such as Ritpanitchajchaval et al. (2021), interpret mountain tourism as adventure tourism motivated by the need for wellness. In this instance, the terms adventure and wellness tourism were utilised interchangeably to characterise the phenomenon. Their central argument revolves around the assertion that mountain tourism prioritises wellness by actively engaging individuals.

In all events, mountain tourism involves tourism activities occurring in specific geographical areas with hills or mountains that possess unique characteristics such as distinctive landscapes, topography, climate, biodiversity and

DOI: 10.4324/9781003392583-11

local communities, which offer visitors a diverse range of outdoor experiences that allow them to connect with nature during their leisure time and immerse themselves in the cultural heritage of the region (Beedie & Hudson, 2003; Statuto & Picuno, 2017).

In Nigeria, inadequate data complicates the understanding of mountain tourism's contribution to the nation's pool of foreign exchange. While local tales and anecdotal evidence in some mountainous communities in Nigeria suggest that mountains are accorded spiritual veneration and significance, not much exists to show whether the development potentials of mountains as tourist destinations are being adequately harnessed. Thus, mountain tourism is not given the requisite official attention despite mountains across Nigeria's geopolitical zones and the associated economic benefits. This neglect is equally reflected in the low level of scientific attention the field has received in the country.

This chapter sets out to explore the landscape of mountain tourism in Nigeria. Methodologically, it relies on a thorough literature review, concentrating on the repercussions of mountain tourism on local mountain communities. Additionally, it investigates the potential and obstacles associated with the development of mountain tourism, leveraging insights from other countries' experiences to illuminate crucial lessons applicable to Nigeria.

Literature Review

People have been travelling to mountains for centuries. While there is an ongoing debate about the origins of scholarly interest in mountain tourism, the literature suggests that journeys to mountains or mountain communities have been a longstanding pursuit driven by various factors such as spiritual, recreational, or exploratory motivations (Apollo & Wengel, 2022; Pfister, 2000; Richins et al., 2016). Mtapuri et al. (2022) argue that the economic advancements and prosperity of the industrial revolution prompted nations worldwide to incorporate tourism into their development strategies. Tourism became an important tool for poverty reduction and economic revival, especially in developing countries (Ashley et al., 2001; Spenceley, 2008). However, these studies seemed to focus more on the phenomenon of the broad field of tourism rather than mountain tourism as a specific aspect of tourism.

A more specific account of the emergence of mountain tourism is found in Beedie and Hudson (2003), where it is argued that scientific attention towards mountain tourism might have been prompted by recreational adventures such as mountaineering sports. The authors note that this form of sports activity was initially the exclusive preserve of the experienced elite (*see also* Apollo, 2017). In a similar line of argument, Mutana and Mukwada (2018) suggest that travelling to mountain areas progressed from aristocrats to scientists and eventually to mountain tourists. From the foregoing, it could be argued that the history of the development of scientific interest in mountain tourism in many areas is characterised by a period of explorations, discoveries, and community interests. These developments and the realisation that mountain tourism constitutes a vital key to economic development (Linca & Toma, 2021) engendered the commodification of mountains; consequently, mountain communities began strategically positioning tourism to drive economic growth. In what follows, a brief reflection on the state of theory on mountain tourism is attempted.

Mountain Tourism: What is the State of Theory?

As a phenomenon, theoretical interest in mountain tourism is relatively recent and under-theorised. It was not until the early 1980s that what seemed like a systematic theorisation on mountain tourism emerged from the work of Dann (1981) when he introduced the idea of motivation as a theoretical framework for tourism. Dan classified the theory into forces of "push" and "pull" and argued that people travel to mountain areas because they are pushed and pulled to travel by some forces/factors that motivate them to move in a particular direction. These forces help to describe how motivation variables push individuals into making travel decisions and how they are pulled or attracted by destination attributes. Uysal and Hagan (1993) later likened push motivations to psychological or emotional aspects, while the pull motivation hypothesis is said to connect to external, situational or (rational) or cognitive aspects (Hanafiah et al., 2010; Yoon & Uysal, 2005). The concepts of "push" and "pull" were subjected to further analysis by Kim and Lee (2002) and Josiam et al. (2005). While Kim and Lee suggest that the attractions in a destination area constitute the pull factors, Josiam et al. submit that push motivations comprise the socio-psychological needs that motivate people to travel. According to Kim and Lee, pull factors include natural attractions, cultural resources, recreational activities, special events, festivals and sundry entertainment opportunities. In summary, the push-pull theory suggests that people travel because they are pushed into making travel decisions by internal psychological forces and pulled by the external forces of the destinations.

It may be helpful to reflect a little on the concept of "push-pull" and its adoption to the explanation of mountain tourism to understand its epistemological implications. The idea of "push-pull" factors is not new in the behavioural sciences. It has been used by early migration scholars such as Ernst Georg Ravenstein in the 1880s to explain how the factors in the places of origin and destination in migratory movements operate to make migrants move out of their place of origin into a new area. What is new about the concept in the current context is its adoption in the field of tourism (which seems to share some affinity with migration and human movement) and the emphasis on the influence of internal socio-psychological forces that act as push factors. A related and more recent pronouncement on the motivation for mountain tourism is the need for self-fulfilment (Taher et al., 2015) and the eudaimonic well-being hypotheses (Ritpanitchajchaval et al., 2023). The locus of the "need for self-fulfilment" and "eudaimonic well-being" hypotheses is that individuals travel to mountain areas for whatever motive (spiritual or otherwise) to enhance their state of well-being and obtain some sense of self-realisation or self-fulfilment, as the case may be. Pfister (2000), in particular, observes that mountain areas are pure and connected to God and represent consecrated images, spirits and legends. The need to have these experiences motivates people, especially hikers, to go to the mountains.

Nonetheless, these ideas (i.e., the need for self-fulfilment and eudaimonic well-being) do not lay any claim to represent a broad and systematised body of theory. Like the push and pull concepts, they offer essential theoretical insights into a field under-theorised in Nigeria. In the next section, a brief excursion is made into Nigeria's historical antecedents of tourism.

Tourism in Nigeria

The year 1472 marked a significant milestone in tourism development in Nigeria. During that period, the country received its first set of Portuguese merchants who arrived in Lagos with a motive for trade and profit. Over the subsequent decades, tourism began to take shape in the country. For instance, around 1962, some concerned Nigerians with a passion for tourism development established the Nigerian Tourism Association (NTA) to promote tourism (Ibeh et al., 2016). The NTA gained recognition from the government and received support and membership in the International Union of Official Travel Organizations (IUOTO) in 1964. The IUOTO later became the World Tourism Organization (WTO). Despite its initial successes, the NTA was unstable and eventually ceased its operations. In 1971, the government engaged the African Development Bank (ADB) to research the potential and viability of tourism in Nigeria. The research affirmed Nigeria's immense potential to become a thriving tourist destination and a significant economic sector (Ibeh et al., 2016).

Consequently, the Nigerian Tourism Board (NTB) was established through Decree No. 54 of 1976 and commenced its operations in 1978. To promote the sector, several conferences were organised to identify and address the obstacles hindering tourism growth in the country. The seventh National Conference on Tourism, held on March 15, 1989, in Maiduguri, Bornu State, highlighted the need to support and restructure the Nigerian Tourism Board (NTB) by commercialising its operations. Following the conference's recommendations, the government approved transforming the Ministry of Trade into the Ministry of Trade and Tourism. These developments reflected Nigeria's commitment to harnessing its tourism potential and positioning itself as an attractive destination for travellers worldwide. Today, there are many tourist sites in Nigeria, such as the Obudu Mountain Resort, Cross River State, Afon Alaye Mountain, Ekiti State, Idanre Hills, and Ondo State.

Mountain Tourism Destinations in Nigeria

Though not naturally endowed with towering mountains, Nigeria offers an array of captivating hilly destinations that can offer tourists from near and far a sense of self-expression and fulfilment. The country's tourism scene showcases the best of its natural and cultural diversity, from natural landscapes and historical landmarks to enchanting waterfalls and cultural gems. Some of the notable tourism sites in Nigeria are discussed below.

Obudu Mountain Resort

Formerly referred to as Obudu Cattle Ranch, the Obudu Mountain Resort is a popular tourist destination in Cross River State, South-South Nigeria. The resort's main attraction is its stunning natural beauty. It offers attractive landscapes with lush greenery, rolling hills, and picturesque valleys. Visitors can enjoy trekking and hiking through the trails, which lead to stunning viewpoints and waterfalls. One of the highlights of Obudu Mountain Resort is the cable car ride or Cable Car Safari, which takes visitors on an exciting journey, offering panoramic views of the surrounding scenery, including the vast grasslands and deep valleys (Ugwukah & Obomamu, 2020).

Olumo Rock

Olumo Rock is located in Abeokuta, Oguns State, Western Nigeria, home to the Yoruba ethnic group. The rock represents a historical and cultural landmark among the Egba people (the mountain community). According to the historical literature of the area, Olumo Rock played a vital role in protecting the Egba people during the Yoruba inter-tribal wars of the 19th century (Anifowose & Olatubosun, 2019; Olaniran, 2023). Visitors to Olumo Rock can explore a network of caves and tunnels carved into the rock. These tunnels served as shelters and defensive positions during times of conflict. Nowadays, these tunnels offer an exciting opportunity for tourists to experience the region's rich history. To reach the top of Olumo Rock, visitors can climb a series of stairs or use an elevator for those who prefer an easier ascent. At the summit, tourists are rewarded with exciting panoramic views of Abeokuta City and its surroundings. Tourism activities around these hills have brought local residents, initially cut off from the rest of the world, into contact with the outside world. Also, local artisans and traders are taking the economic advantage of tourism by investing in small trades around the base of the rock, offering souvenirs, arts and crafts, thereby adding value to the local economy (Ugwukah & Obomamu, 2020).

Erin Ijesha Waterfalls and Idanre Hills

The Erin Ijesha Waterfalls, also known as Olumirin Waterfalls, is a natural wonder in Erin-Ijesha, Osun State, Nigeria. The waterfall is a series of seven cascades that flow down the rocks, creating picturesque scenes and refreshing natural pools. Visitors to Erin Ijesha Waterfalls can hike through lush greenery and rocky terrain to reach the different levels of the falls. Each level offers a unique and enchanting view of the cascading water; some have natural pools. Erin Ijesha Waterfalls has become increasingly popular as a tourist attraction, drawing both local and international visitors seeking a peaceful and refreshing escape from city life.

Another mountain feature with cultural and symbolic significance is Idanre Hills. Located in Idanre Town, Ondo State, Idanre Hill is a cluster of picturesque hills and valleys. The hills have historical and cultural significance as they are home to ancient structures, artefacts, and inscriptions providing insights into the region's past and social history (Ale et al., 2020; Ikusemiju & Osinubi, 2020; Jeje et al., 2023). Tourists can explore the hills by climbing a series of steps and paths that lead to different historical sites and scenic viewpoints. These hills also serve as important spiritual centres for residents and citizens who ascribe celestial importance to the hills and seek spiritual solutions and answers to their problems.

Potential of Mountain Tourism Development in Nigeria

The potential of mountain tourism to stimulate socio-economic development at the country and local community levels is already documented (Briedenhann & Wickens, 2004; Frechtling & Hara, 2016; Pfister, 2000; Taher et al., 2015). Linca and Toma (2021) argue that the significance of mountain tourism extends beyond the national economic benefits because it impacts the well-being of local communities and enriches tourists' experiences. Evidence from other parts of the world shows that mountain tourism has the potential to boost local economies and improve the livelihood outcomes of communities in mountain regions such as the Nepalese Himalayas, Mount Everest (Apollo, 2017) and Mount Kenya (Chiyumba, 2015), as well as Malaysia (Taher et al., 2015). In Malaysia, for instance, the realisation of the potential of mountain tourism to attract high yield tourists contributing to the country's GDP was the basis for Malaysia's integration of mountain tourism into the 10th Malaysia Plan and Tourism National Key Economic Area (NKEA) and the Tourist Transformation Plan (Taher et al., 2015). Today, Malaysia remains one of the emerging mountain tourism destinations and has made the sub-sector a significant policy focus in the drive towards greater local and international participation. The experience of Nepal is another example. According to Beedie et al. (2003), the setting of the Himalayan Mountain in Nepal has been commodified with strict administrative control to generate revenue to run the economy. Despite Nepal's dismal economic situation, the host community still benefits socioeconomically from mountain tourism.

The previous country-based analysis proposes that Nigeria, given its unique mountainous properties, can revive the economy and revitalise the economy of rural communities, create jobs in rural mountain communities and inspire rural employment. Furthermore, this should be done in the interest of local community residents (Nepal & Chipeniuk, 2005). Through the urbanisation of mountain spaces, mountain tourism can directly benefit the communities by generating employment opportunities and improving the overall quality of life (Briedenhann & Wickens, 2004).

Prospects and Challenges of Mountain Tourism Development in Nigeria

Mountain tourism can impact local communities in several ways, including a shift in value systems as residents come in contact with foreigners (Eneji et al., 2016; Gnanapala & Sandaruwani, 2016). Furthermore, mountain tourism can be a feasible option for developing and reducing poverty and unemployment in mountain communities in many less developed countries afflicted by rural poverty, such as Nigeria. With pristine mountain objects in some parts of Nigeria, the country can make significant gains from the tourism subsector, transforming the poor condition of mountain communities and boosting the national economy. For instance, the increase in tourist visits and patronage at Obudu Mountain Resort has impacted the host community positively. The immediate host community (Becheeve community) was initially made up primarily of farmers and hunters. With the increase in tourist activities around the community, many community members have become engaged in tourism-related activities such as jobs in hotels, sales of food, drinks and other local products, transportation and tour-guiding services. Tourism within this area has also supplied essential amenities, including schools, electricity, security, accessible roads, and financial and communication facilities (Amalu et al., 2012). This shows how mountain tourism activities can stimulate a shift in the socio-economic status of mountainous communities.

Nonetheless, there are indications that some mountain communities do not benefit sufficiently from the impacts of mountain tourism activities. For example, Amalu and Ajake (2012) examined the employment structure of Obudu Mountain Resort in Cross River State in the tourism industry. The authors focused on the hotels within the resort. They observed that, despite the high rate of tourism activities in the study area, residents in the community were yet to have the full socio-economic benefits of tourism in terms of employment, infrastructural development, and adequate improvement in their standard of living. Similarly, Uwadiegwu (2002) revealed that local people's participation in tourism development in Nigeria is yet to be given adequate attention. The author notes that the people of Becheeve (host of Obudu Mountain Resort), for instance, have been denied the right to maintain and control their cultural heritage and manage the positive and negative impacts of tourism activities. This reinforces the view prevalent in the literature on tourism that local communities are often sidelined by governments in the management of mountain tourism spaces in developing countries (Ajake et al., 2010).

Remarkably, despite global efforts towards mountain tourism development, many regions in Nigeria have yet to harness the full socio-economic potential of the mountain tourism sub-sector. A north-south comparison of tourism activities in Nigeria revealed that most of the visitors in 2018 were in the southern part of the country (Amalu et al., 2020). This difference in regional tourism activities between the north and south, according to Amalu et al., can be attributed to the poor destination image of the north, which is most affected by insecurity plaguing the country. The problems of insecurity, high levels of corruption, inadequate infrastructure and rampant kidnapping in Nigeria pose a significant challenge to mountain tourism development in the country. Most visitors to mountain sites and hills that attract essential economic benefits to Nigeria and mountain communities are foreigners. The frequency with which foreigners are kidnapped for ransom in Nigeria, coupled with the country's battered image, makes the country diminish in value as a tourist destination.

Conclusion

Mountain tourism has been acknowledged globally as an essential sector capable of driving socio-economic change and development, particularly in the global south. The experiences of countries such as Malaysia and Nepal demonstrate this fact. Despite the vast mountain features throughout Nigeria, the country has not been able to harness the full potential of the tourism subsector. Several factors have been pointed out as militating against the growth and potential of mountain tourism in Nigeria. Some of them include the neglect and exclusion of mountain communities in the management of mountain tourism, widespread institutional corruption, insecurity, and lack of government will. The chapter observed that, although mountain tourism appeared to positively impact mountainous communities, much needs to be done to maximise the benefits of tourism activities in mountain communities.

Management Implications

The evidence is clear that mountain tourism can contribute greatly to national development globally. Nigeria can benefit tremendously from its vast mountain features and endowment by mobilising the political will and appropriate policy frameworks. This chapter puts forward the following recommendations with policy implications for mountain tourism management in the country.

- Incorporating mountain communities into the management of mountain tourism activities is crucial. By doing so, trust between the government and the people can be strengthened, ultimately fostering sustainable communities. This alignment with Sustainable Development Goal (SDG) 11 enhances global efforts towards achieving sustainable development objectives.
- Mountain tourism's socio-economic benefits can only be maximised without insecurity and conflicts. Efforts should be intensified by governments at all levels to create an environment that is secure for mountain

tourism to thrive. The reports of incessant kidnapping and the activities of bandits and herders across the country do not augur well for mountain tourism in the country.

- Efforts should be made to support studies in the field of mountain tourism. There is a lack of adequate data in the sub-sector for policy formulation, implementation and management of the sub-sector.
- The government should make mountain tourism a significant policy focus and integrate the sub-sector into national development plans. Mountain tourism should be a key economic area with adequate policies to regulate its management.

References

Ajake, A. O., Egbe, C. A., Ambo, M., Emanghe, E., & Eneji, C. O. (2010). Impact of conservation activities of non-government organisation on sustainable tourism and economic growth in Cross River State, Nigeria. *Journal of humanities and development studies*, 1(1), 21–32.

Ale, T. A., Alade, J. M., & Ogunruku, M. P. (2020). Harnessing Natural and Cultural Resources for Recreational Purpose in Boosting Tourism Development in Ondo State, Nigeria. *International Journal of Research and Analytical Reviews*, 7(3), 747–782.

Amalu, T. E., & Ajake, A. O. (2012). An assessment of the employment structure in the tourism industry of Obudu Mountain resort, Cross River state, Nigeria. *Geo-Journal of Tourism and Geosites*, 9(1), 35–56.

Amalu, T. E., Okpara, D. E., Omeje, V. U., Obi, P. O., & Okoruwa, D. E. (2012). Socio-economic implication of tourists' appraisal of the Obudu Mountain resort, Cross River State, Nigeria. *Journal of Environmental Science, Toxicology and Food Technology*, 1(3), 37–44.

Amalu, T., Phil-Eze, P., & Ajake, A. (2020). Assessing the impact of economic and cultural diversity on tourism development in Nigeria. *GeoJournal*, 85, 1457–1468. https://doi.org/10.1007/s10708-019-10032-2.

Anifowose, T., & Olatubosun, J. (2019). Historic Places and Buildings in Yoruba land. *International Journal of Advances in Scientific Research and Engineering* , 5(6), 79–84.

Apollo, M. (2017). The true accessibility of mountaineering: The case of the High Himalaya. *Journal of Outdoor Recreation and Tourism*, 17, 29–43. https://doi.org/10.1016/j.jort.2016.12.001.

Apollo, M. & Wengel, Y. (2022). *Mountain tourism: A critical perspective*. Routledge. https://doi.org/10.4324/9781003095323.

Ashley, C., Roe, D., & Goodwin, H. (2001). *Pro-poor Tourism Strategies: Making Tourism Work for the Poor. A Review of Experience*. Overseas Development Institute.

Beedie, P., & Hudson, S. (2003). Emergence of mountain-based adventure tourism. *Annals of tourism research*, 30(3), 625–643. https://doi.org/10.1016/S0160-7383(03)00043-4.

Birendra, K. C., Dhungana, A., & Dangi, T. B. (2021). Tourism and the sustainable development goals: Stakeholders' perspectives from Nepal. *Tourism Management Perspectives*, 38, 100822. https://doi.org/10.1016/j.tmp.2021.100822.

Briedenhann, J., & Wickens, E. (2004). Tourism routes as a tool for the economic development of rural areas vibrant hope or impossible dream? *Tourism Management*, 25(1), 71–79. https://doi.org/10.1016/S0261-5177(03)00063-3.

Chiyumba, A. A. M. (2015). *Mountain tourism and its contribution to development in western Mt. Kenya Region: an assessment using the value chain approach* (Doctoral dissertation).

Dann, G. M. (1981). Tourism motivations: An appraisal. *Annals of Tourism Research*. 8(2), 189–219.

Eneji, M. A., Odey, F. A., & Bullus, M. L. (2016). Diversification of Nigeria's economy; impact of tourism on sustainable development in Nigeria. *International Journal of Research in Humanities and Social Studies*, 3(5), 36–44.

Frechtling, D. C., & Hara, T. (2016). State of the world's tourism statistics and what to do about it. *Tourism Economics*, 22(5), 995–1013.

Gnanapala, W. A., & Sandaruwani, J. A. R. C. (2016). Socio-economic impacts of tourism development and their implications on local communities. *International Journal of Economics and Business Administration*, 2(5), 59–67.

Hanafiah, M.H.M. & Harun, W. F. M. (2010). Tourism demand in Malaysia: A cross-sectional time series analysis. *International Journal of Trade, Economics & Finance*, 1 (1), 80–83.

Ibeh, P. E., Ikegwu, E. M., & Amaechi, C. J. (2016). Tourism development: a revolutionalised sector as a panacea for dwindling hospitality business in Nigeria. *International Journal of Science and Society Yabatech* 4(1), 39–46.

Ikusemiju, T. M., & Osinubi, O. B. (2020). Improving the cultural and historical tourism resources for sustainable development in Ondo state–a survey of Idanre hills and resort centre, Idanre. *Nigerian Journal of Environmental Sciences and Technology*, 4(1), 197–203.

Jeje, L. K., Orimoogunje, O., & Olusola, A. (2023). Hills and ridges in southwestern Nigeria. In A. Faniran, L. K. Jeje, O. A. Fashae, & A. O. Olusola (Eds.), *Landscapes and Landforms of Nigeria* (71–87). Springer. https://doi.org/10.1007/978-3-031-17972-3_5.

Josiam, B.M., Kingsley, T.R. & Kim, Y.K. (2005). Involvement and the tourist shopper: Using the involvement construct to segment the American tourist shopper at the mall. *Journal of Vacation Marketing*, 11(2), 135–154. https://doi.org/10.1177/135676670505257.

Kim, S. & Lee, C. (2002). Push and pull relationships. *Annals of Tourism Research*, 29(1), 257–260.

Linca, A., & Toma, E. (2021). Study regarding the evolution of mountain tourism and rural mountain tourism in Romanian Carpathians. *Scientific Papers Series Management, Economic Engineering in Agriculture and Rural Development*, 21(1), 463–470.

Mtapuri, O., Camilleri, M. A., & Dłużewska, A. (2022). Advancing community-based tourism approaches for the sustainable development of destinations. *Sustainable Development*, 30(3), 423–432.

Mutana, S., & Mukwada, G. (2018). Mountain-route tourism and sustainability. A discourse analysis of literature and possible future research. *Journal of Outdoor Recreation, Tourism and Research Planning*, 24, 59–65. https://doi.org/10.1007/s11629-018-4771-7.

Nepal, S. K., & Chipeniuk, R. (2005). Mountain tourism: Toward a conceptual framework. *Tourism Geographies*, 7, 313–333. https://doi.org/10.1080/14616680500358595.

Olaniran, H. D. (2023). Urban Geoheritage Site: The Example of Olumo Rock in Abeokuta City, Ogun State, Nigeria. In A. Faniran, L. K. Jeje, O. A. Fashae & A. O. Olusola (Eds.), *Landscapes and Landforms of Nigeria* (267–278). Springer.

Pfister, R. E. (2000). Mountain culture as a tourism resource: aboriginal views on the privileges of storytelling. In P. M. Godde, M. F. Price & F. M. Zimmermann (Eds.), *Tourism and development in mountain regions* (115–136). CABI.

Rasulovich, K. A. (2021). The role of agro-tourism in the development of socio-economic infrastructure in rural areas. *Наука и ёбразёвание сегёдня*, 3(62), 13–14.

Richins, H., Johnsen, S., & Hull, J. S. (2016). Overview of mountain tourism: substantive nature, historical context, areas of focus. In H. Richins & J. S. Hull (Eds.), *Mountain tourism: Experiences, communities, environments and sustainable futures* (1–12). CABI.

Ritpanitchajchaval, N., Ashton, A. S. & Apollo, M. (2023). Eudaimonic well-being development: Motives during mountain-based adventure tourism. *Journal of Outdoor Recreation and Tourism*, 42, 100607. https://doi.org/10.1016/j.jort.2023.100607.

Spenceley, A. (2008). *Responsible Tourism: Critical Issues for Conservation and Development*. Earthscan.

Statuto, D., & Picuno, P. (2017). Valorisation of vernacular farm buildings for the sustainable development of rural tourism in mountain areas of the Adriatic-Ionian macro-region. *Journal of Agricultural Engineering*, 48, 21–26. https://doi.org/10.4081/jae.2017.603.

Taher, S. H., Jamal, S. A., & Sumarjan, N. (2015). A survey of hikers characteristics at Prominand Mountains in Malaysia. In Radzi, S. M., Sumarjan, N., Chik, C. T., Zahari, M. S. M., Mohi, Z., Bakhtiar, M. F. S., & Anuar, F. I. (Eds.), *Theory and Practice in Hospitality and Tourism Research* (213–220). CRC Press.

Ugwukah, A., & Obomanu, V. S. O. (2020). Tourism and Economic Development in Nigeria: A Critical Assessment of the Obudu Mountain Resort, Cross River State 2008 to 2018. *Journal of Tourism, Hospitality and Sports*, 80, 2312–5187.

Uwadiegwu, B., O., (2002), *Basics of recreation and tourism facilities, planning*. Enugu: Academic Publishers.

Uysal, M., & Hagan, L. (1993). Motivations of pleasure travel and tourism. In M. Khan, M. Olsen, & T. Var (Eds.), *Encyclopedia of Hospitality and Tourism* (798–810). Van Nostrand Reinhold.

Yoon, Y. & Uysal, M. (2005). An examination of the effects of motivation and satisfaction on destination loyalty: A structural model. *Tourism Management*, 26, 45–56. https://doi.org/10.1016/j.tourman.2003.08.016.

10 Leveraging Pro-Poor Tourism for Poverty Alleviation in the Eastern Highlands of Zimbabwe

Tendai Chibaya and Zibanai Zhou

Introduction

Tourism has morphed into a multi-trillion-dollar sector, with countries worldwide embracing it inspired by its employment creation, foreign currency generative capacity, and poverty alleviation prowess (Chibaya & Zhou, 2023; Musavengane, 2018; Zhou, 2018). Although COVID-19 induced downward revisions of international tourism traffic, the desire to visit unspoilt rural tourist spaces and mountain resorts is still increasing (Lak & Khairabadi, 2022; Apaza-Panca et al., 2020). This creates vast opportunities for marginalised communities to tap into the tourism boom. Shifting global trends towards heightened tourist interest in mountain areas has renewed hope for developing countries like Zimbabwe, which boasts of such globally acclaimed destinations as the Eastern Highlands. However, concern remains that the socio-economic benefits of tourism are unequal and continue to elude local communities, particularly in remote mountainous areas which are characterised by peculiar geographical challenges (Chipfuva 2021; Scheyvens, 2007). Most rural communities located in mountainous areas continue to grapple with chronic challenges such as poor access to markets due to dilapidated roads, recurrent droughts and flooding due to climate change, pillaging of local resources typically by foreign companies, high poverty levels, and general under-development in multiple respects (Chipfuva, 2021; Mutana et al., 2013; Scheyvens 2007). Such challenges persist even while mountain tourism is hailed as having the potential to improve infrastructure and transport and to bring substantial economic development and social change to mountain regions and their surrounding communities (UNWTO, 2018).

The Eastern Highlands communities in Zimbabwe grapple with numerous tragedies, such as cyclones and geographical challenges to agriculture. However, the area is popular with regional and international tourists who want to hike, among other things. Its natural and mountainous breath-taking scenery (for example, in the Bvumba, Chimanimani and Inyangani mountains), its abundance of wildlife species in Chimanimani National Park and Nyanga National Park, and mountain trekking attract tourists from all walks of life. However, due to the abundance of such a wide range of natural tourist

DOI: 10.4324/9781003392583-12

attractions, local communities have yet to benefit meaningfully from commercially exploiting such resources. While mountain tourism continues to thrive, adjacent communities grapple with high poverty levels. However, tourism has been hailed as a panacea to challenges confronting most remote communities, such as high rural-urban migration, uninspiring infrastructural facilities, marginalisation, endemic poverty levels, limited employment opportunities, and a general lack of development, poor access to potable water, inadequate basic health, sanitation and education facilities, poor network coverage, hence poor access to information among others. The study is significant in that it sheds light on most topical issues around the advancement of local communities and equitable development through the use of local resources found in communities. However, limited studies have examined such issues in the context of pro-poor mountain tourism in Zimbabwe. The study provides insights into how local communities can benefit from tourism resources in their area.

Theoretically, the study addresses the under-representation of pro-poor mountain tourism studies in the contemporary Zimbabwean tourism discourse. In addition, the study would assist tourism policymakers in implementing devolution at a community level in the tourism sector, which would go a long way toward benefiting local communities and reducing leakages. The objectives of the chapter are outlined as follows:

- examine the socio-economic and ecological benefits of pro-poor mountain tourism to communities in the Eastern Highlands,
- analyse the challenges which inhibit local communities from fully benefiting from pro-poor mountain tourism, and
- determine pathways through which local communities optimise pro-poor mountain tourism to derive maximum benefits.

The chapter unfolds as follows: a background follows the introduction. Subsequently, the methodology is provided. Then, the results and discussion sections follow. Finally, a conclusion is given, accompanied by management implications.

Background

The Concepts of Mountain Tourism and Pro-Poor Tourism

Mountain tourism is one of the tourism segments in which tourism activities are experienced in a mountain environment (Apollo & Wengel, 2022). Mountaineering tourism is divided into three groups: trekking, climbing, and hiking (Apollo, 2021). Tourists travel to mountain areas for adventure experiences, while others are motivated to visit holy mountain places. However, this particular type of tourism is changing from being an elite sport to becoming more of mass tourism as a result of tourism commodification, commercialisation, and diversification (Beedie & Hudson, 2003).

Pro-poor tourism is an approach that promotes and generates more significant benefits for low-income people (Pro-Poor Partnership, 2004). It is concerned with how tourism affects the socio-economic livelihoods of communities at a local level (Musasa & Mago, 2014; Rogerson, 2006). In line with mountain tourism, Apollo and Wengel (2022) opine that mountains, directly and indirectly, support the livelihood and well-being of billions of people who live in mountain areas and below with intangible (recreation and education opportunities) and tangible (food, water, and energy) resources. The UNWTO (2018) outlines that as mountain tourism becomes popular, it will be associated with multiple benefits, including economic development, improved transportation and infrastructure, and expansion of social services. Such development benefits mountain areas and surrounding communities. In developing countries, mountain tourism is significant as it complements traditional sources of livelihoods such as agriculture.

Consequently, pro-poor mountain tourism has thus been widely promoted and accepted as a panacea for poverty alleviation. Mountain tourism can also be used as a vehicle for the preservation and maintenance of the local people's culture and heritage. The local people can educate foreign tourists about their culture through their lifestyle, clothing, and language (Apollo & Wengel, 2022). However, despite the positive socio-economic benefits of mountain tourism, local people still grapple with challenges such as rising cost of living, noise, traffic congestion and pollution, changes in lifestyles, dressing, and local language (Apollo & Andreychouk, 2022; Apollo & Wengel, 2022). For example, when the European Alps mountain region was overbuilt and overdeveloped to support mountain tourism in the 1960s and 1980s, it lost its cultural attractiveness/allure (Price et al., 1999).

Furthermore, mountain tourism development improves infrastructure, income, employment opportunities, and the preservation of natural resources. Despite the positive benefits associated with mountain tourism, tourists can put pressure on water, land use, air, flora, and fauna (Apollo, 2021). According to Apollo (2021), mountain tourism activities can damage paths and trails. Wall and Mathieson (2006) assert that mountain tourism causes soil erosion. It is further highlighted that access roads and accommodation construction, excessive dumping of garbage from mountain tourists, and careless use of fire in the mountains destroy fauna and flora (Wall & Mathieson, 2006). Similarly, camping, off-road motor vehicles, trampling, mountain bike riding, and horse riding negatively affect vegetation and soil.

However, the concept of mountain tourism (closely related to pro-poor tourism) remains underexplored in Zimbabwe, more so in unison with poverty alleviation. Although empirical studies on pro-poor tourism are available (Chipfuva, 2021; Musavengane, 2018), very few have analysed the impact of pro-poor mountain tourism on poverty alleviation. The study examines pathways through which local communities can draw maximum benefits from mountain tourism, how to reduce leakages commonly associated with mountain tourism, and how to enhance the wider circulation of mountain tourism income within the local economy. Tourism development as a poverty reduction strategy has long been an essential area of research in developing countries; however, limited studies have examined this aspect in the context of mountain tourism. Most studies are predominantly macro or national context with limited studies at a local level.

Research suggests that poverty alleviation through tourism development is not automatic in light of the disproportionate distribution of resources and non-involvement of local communities in tourism development decision-making processes. Scheyvens (2007) maintains that a geographical overlap always exists between areas of high tourism activity and poverty. Manjengwa et al. (2016) posit that poverty increased in Zimbabwe due to economic structural adjustment programmes adopted in the 1990s. The Eastern Highlands communities of Zimbabwe rely heavily on handouts from the non-governmental (NGO) sector, which is evidence of extreme poverty (Kinsey, 2010. A similar trend characterises most rural communities in Zimbabwe, which grapple with widespread poverty and high unemployment rates.

In support of the above, many protagonists of pro-poor tourism believe that tourism's potential for poverty alleviation is more significant than other economic sectors because of its size, labour intensity, and the many opportunities for business linkages it presents. However, the premise is opposed by the argument that tourism is frequently led by the private sector, causing high leakages (money flowing out of the area/country), fewer linkages, loss of access to natural resources, as well as cultural and social disruption. To this end, Chipfuva (2021) sought to develop a pro-poor tourism framework for Zimbabwe using the case of Chitomborwizi East small-scale commercial farms in Makonde district in Mashonaland West province. The study found that agriculture was the main income-generating activity for communities; however, poverty worsens due to low rainfall, frequent drought, and high input costs.

In yet another study, Musavengane (2018) examined strategies for pro-poor tourism development in Zimbabwe to enhance local economic development and alleviate poverty in poor communities using documentary analysis. The study established the need for Zimbabwe to enhance economic opportunities for pro-poor actors to pursue tourism ventures and obtain a fair livelihood. This aligns with the United Nations Sustainable Development (UNSD) Goal 8 of decent work and economic growth (UN, 2017).

In addition, Gohori and van der Merwe (2022a) examined the potential of tourism to empower local communities in Manicaland province. Furthermore, Gohori and van der Merwe (2020) sought to develop a tourism-community-development framework. Similarly, Gohori and van der Merwe (2022b) examined the perspectives of local communities towards the limitations of community participation in tourism in Manicaland province. Mutana et al. (2013) assessed the contribution of tourism operators towards pro-poor tourism development in Zimbabwe, focusing on communities surrounding Victoria Falls and Hwange. The study found that tourism operators were willing to develop pro-poor tourism but were impeded by political and economic challenges. Despite the extensive research on pro-poor tourism and poverty alleviation, findings remain mixed globally.

In Zimbabwe, 80 per cent of the population resides in rural areas. Rural areas lack economic development and essential social services, and have limited employment opportunities and limited access to devolution funds compared to urban areas. In addition, rural communities are characterised by rural-urban migration, poor mobile network services, dilapidated roads and transport systems, and high dependence on NGOs. Tourism development becomes significant in this context as it can address some of these challenges. Rural tourism development in Zimbabwe has been undertaken within the Communal Areas Management Programme for Indigenous Resources (CAMPFIRE) framework, whose effect on poverty reduction has been marginal (Logan & Mosely, 2002. Lurdi and Bird (2007) posit that poverty comes in various forms, which include lack of income, limited access to productive resources sufficient to ensure sustainable livelihood, hunger and malnutrition, ill health, or lack of access to education and other essential services. Increased morbidity and mortality from illness, homelessness and inadequate housing, unsafe environments, social discrimination, and exclusion from decision-making on development are other facets of poverty. Poverty is associated with rurality. Kinsey (2010) notes that poverty in Zimbabwe is overwhelmingly rural, yet rural areas have significant tourist attractions that can be leveraged to reduce poverty. Consequently, tapping into such tourism resources would go a long way to improving the socio-economic livelihoods of mountain community areas.

Pro-poor tourism is a form of tourism that benefits people with low incomes (Scheyvens, 2007). Pro-poor tourism was developed in response to failed mainstream development themes such as modernisation and neo-liberalism which exacerbated poverty and inequalities. Tourism earns foreign currency, thus alleviating poverty (Spenceley & Meyer, 2012). Scheyvens (2014) outlines that local people in areas endowed with abundant natural resources often lack access to basic infrastructure, sanitation, health facilities, and schools, while tourists have access to all amenities. To this end, mainstream tourism is criticised based on its insignificant impact on poverty alleviation, attributed to foreign domination, high-income leakages, and fewer linkages with the local economy.

Social Capital

Social capital denotes rules, norms, obligations, reciprocity, and trust embedded in social relations, social structures, networks, and society's institutional arrangements, enabling individuals and community members to achieve objectives through solidarity (Moscardo, 2015). It entails networks, connectedness, and group membership, which increase people's trust and ability to work together to access resources. Social capital determines mountain tourism's success and addresses inequalities in the distribution of mountain tourism benefits (Borinodo & Coromina, 2018). Moscardo et al. (2017) outline that social capital is a source of power, influence, and dissipation of information, and fosters solidarity among community networks. Communities with a high level of social capital tend to succeed in tourism development (Moscardo et al., 2017)

Methodology

The Study Area: Zimbabwe's Eastern Highlands

The research was based on the Eastern Highlands of Zimbabwe. The Eastern Highlands comprise a mountain range on the border of Mozambique and Zimbabwe. There are three mountain groups: Chimanimani, Nyanga, and the Bvumba Mountains. The area extends south and north for about 300 kilometres through Manicaland province in Zimbabwe and Manica province in Mozambique.

Incredible mountain ranges, rolling mists, cultivated forests, apple orchards, vast tea estates, and spectacular waterfalls characterise the Eastern Highlands. The area is very popular with outdoor tourist enthusiasts participating in hiking, camping, birding, fishing, and golf tourism. The Eastern Highlands are characterised by microclimates, which are influenced by the enormous mountains and rolling valleys. Consequently, the flora and fauna in the area are unique.

Research Design

The study adopted a qualitative approach. Snowball sampling was used to select local community representatives who were not easily accessible. The residents lived in sparsely populated areas where it was difficult to identify potential respondents; hence, there was a need to seek guidance from the first respondents. Through snowball sampling, eleven respondents representing communities were interviewed. This was accompanied by a purposive sampling that selected representatives from the Zimbabwe Tourism Authority (ZTA) (1), ZimParks (1), and the Environmental Management Agency (EMA) (3). Overall, sixteen face-to-face interviews were conducted. Interviews were transcribed and subsequently analysed thematically per Walter's (2016) framework. The framework includes data familiarisation, coding, consolidating codes into basic themes, generating organised themes, establishing the networks, and describing and analysing the network.

Findings

Socio-economic and ecological benefits of tourism on Eastern Highlands local communities

Preservation of a Clean and Beautiful Environment

Findings showed that adhering to the stipulated number of tourists on tourist sites goes a long way to preserving a clean and beautiful environment in the Eastern Highlands. This is based on the fact that crowded tourist sites are usually associated with littering, dumping of plastic and other non-biodegradable material that are detrimental to wildlife, toxic to flora, and dangerous to human life. The following excerpt illustrates this:

> The preservation of a clean and beautiful environment is one of the core responsibilities of my department as I am employed by the Environment Management Agency of Zimbabwe. As a nation, we are working on the motto, "Zero tolerance to litter: my environment, my pride". All actors should make sure that they take part in the national clean-up day that is set aside on the first Friday of each month.
>
> (Participant, 1)

This implies that all stakeholders should make sure that the environment is clean for the benefit of present and future generations in mountain regions and surrounding areas.

Promoting the Socio-Economic Interests of Mountain Tourism Stakeholders (Economic and Other Interests of Local Mountain Dwellers)

Most respondents highlighted various ways through which the economic interests of mountain tourism stakeholders can be promoted. The various stakeholders encompassed the government, local authorities, mountain dwellers, and the ZTA. It was stated that local control of mountain tourism ensures that government gets optimum taxes and foreign currency. The foreign currency revenue is most significant to the Zimbabwe Parks and Wildlife Management Authority (Zimparks) as the custodian of wildlife resources. Zimparks needs financial resources to support various wildlife conservation initiatives and programmes. Furthermore, for mountain dwellers, it means income is ploughed back into the local economy, thereby boosting aggregate demand for products and other goods and creating more employment opportunities through upward and downstream linkages.

In the Eastern Highlands, many independent cultural dance groups complement mountain tourism. The local traditional leaders also benefit from the preservation of local cultural values, thus guarding against the commercialisation of the culture. The *Ndau* and *Manyika* people have a rich cultural history, dances, and folklore. However, too many dancing troupes are competing for the same tourist market. The notable setback is that such groups outcompete customers and end up charging below-market rates to lure customers. This reduces potential income. In addition, some groups end up staging cultural dances, which dilutes the authenticity of such dances. This may negatively affect tourist arrivals in mountain areas because the authenticity of the cultural tourism offerings initially attracts mountain tourists. A formally organised community cultural dance group is better positioned to charge more and earn more profits than many separate and competing troupes. This would preserve the culture and increase income earned.

Dominance of Foreign Companies and Other Remote Actors in the Mountain Tourism Value Chain

Further data analysis confirmed the dominance of large foreign companies in the Eastern Highlands mountain tourism value chain, which prevents local communities from benefiting meaningfully from mountain tourism. It was outlined that the Eastern Highlands are primarily promoted as little England following its weather patterns and climatic conditions, which resemble England. As such, tour operators (TOs) and travel agencies (TAs) based in Western markets dominate the composition and promotion of mountain tourism packages, bypassing locally based intermediaries. As a result, prospective customers pay for mountain holiday vacations outside Zimbabwe, thus exacerbating leakages. Moreover, other companies are based in major cities, particularly Harare and Mutare, which intercept potential mountain tourism business; hence, local companies located in the Eastern Highlands remain excluded. Consequently, mountain tourism proceeds rarely trickle down to the mountain dwellers or the local economy as their benefits flow to major cities or outside Zimbabwe.

Foreign Ownership in the Accommodation Service Provision

Additionally, a similar trend was noticed in the provision of accommodation services, whereby the subsector is dominated by large chain groups that crowd out small to medium enterprises (SMEs) and other locally owned facilities. This causes intense competition among accommodation service providers for customers, forcing them to offer discounts and, therefore, cutting revenue margins just to break even. However, there is scope for local communities to get maximum profits through collaboration, cooperation and community cooperatives instead of competing. The following narrative illustrates this:

> This is an opportunity for us local communities to be entrepreneurs through collaboration and pooling of financial and human resources.
>
> (Participant, 4)

Chain groups account for a large share of the market. SMEs have some success in providing the same service, albeit on a small scale. Offering the same service or product by many SMEs erodes profit margins, and eventually, locally-owned facilities lose out to big chain groups that offer customised mountain tourism products. It was also established that market inefficiencies characterise the mountain tourism market in the Eastern

Highlands as tourism operators spread resources thinly, compromising service delivery. Local mountain tourism communities must leverage social capital to form diversified companies that can handle all ground services, advertising, accommodation, meals, entertainment, activities, and targeted marketing. This would position tourism operators to provide quality mountain tourism experiences and charge premium rates, thus increasing revenue. Diversified companies are better placed to compete with established chain groups.

Additionally, diversified companies offer multiple backward and forward integration opportunities, thus enabling local tourism operators to control the mountain tourism value chain. This goes a long way to guaranteeing value for money, customer satisfaction, positive word of mouth, repeat purchases preserving the environment, and substantial profit margins. High profit margins entail that the national economy benefits through increased taxes and local authorities from greater levies.

Excessive Income Leakages

Findings further showed excessive leakages in the Eastern Highlands mountain tourism value chain. This is attributed to the dominance of foreign tourism and hospitality corporations such as hotel chains, tour operators, and travel agents who externalise revenue outside communities. This issue is of significant concern as it weakens the dominant view that mountain tourism alleviates poverty. The following narratives illustrate this:

> Most hospitality organisations in the Eastern Highlands are owned by external business people who live in the capital city of Harare and other foreign countries. The local communities do not have shares in these businesses. Local communities' benefits are only limited to the selling of produce to tourists and producers at local markets as well as employment (Participant, 10).

In the case of the Eastern highlands, the ownership at the national level creates intra-national inequalities, which might lead to the creation of local elites who would be as exploitative as multinationals.

Social capital

Mountain dwellers can leverage community social capital to derive maximum benefits from mountain tourism in the Eastern highlands. Social capital might be witnessed through relations or networks among individuals or collectives (Häuberer, 2011). This was supported by one of the participants who said:

> For the Zimbabwean mountain tourism to be sustainable, there is a need to forge a strong alliance network from different stakeholders to achieve a

> common goal. If we remain fragmented, then meaningful benefits from mountain tourism will remain elusive
>
> (Participant, 11)

If stakeholders leverage social capital as one of the core functional resources, it drives other resources such as human and financial capital into fruitful action.

Financial resources

Financial capital entails financial resources available to individual mountain dwellers and at the household level. This includes savings, credit supplies, and regular remittances such as pensions, which present opportunities for pursuing different livelihood options (Carney 1999). However, the tragedy with rural communities is that they have been permanently excluded from gaining access to financial services because commercial banks view investment in rural areas as non-viable (Makoni, 2014; Zhou & Chibaya, 2022). This prevents mountain dwellers from pooling resources and establishing diversified companies. Financial resources are a vital asset for mountain dwellers as they unlock potential investment opportunities in mountain tourism. Mountain dwellers with financial resources can invest in information and communication technologies (ICTs), establish diversified companies, package mountain tourism products, build hotels and lodges, and buy coaches to boost mountain tourism. Income obtained from successful mountain tourism activities can then be used to diversify the activities in the mountain tourism value chain.

Challenges inhibiting mountain dwellers from benefiting from mountain tourism

Participants cited diverse challenges that prevent the Eastern Highlands dwellers from benefiting from mountain tourism. These were outlined as lack of financial resources, dominance of foreign companies, excessive leakages, the fragmented nature of small to medium tourism enterprises, lack of involvement of local communities in planning, and limited education on how to reduce causes of climate change and of destruction of biodiversity from other stakeholders through wild fires and clearing of land for construction of buildings.

Nature of jobs and sustainability

Findings show that jobs available to local communities in the Eastern Highlands are largely menial and of low rank, with their sustainability uncertain on account of the seasonal nature of mountain tourism activities.

Optimisation of pro-poor mountain tourism to address poverty issues

Most participants believed that mountain tourism can be optimised by reconfiguring the ownership aspect of mountain tourism value chain facilities to ensure that they are firmly in the hands of local people. In addition, there is a need to reorient mountain tourism activities to be appropriately aligned with the local communities' sources of livelihood and to retain revenue within the local economy.

Discussion

The study sought to examine the socio-economic and ecological benefits of mountain tourism in the Eastern Highlands, to analyse the challenges preventing local communities from benefiting fully from pro-poor mountain tourism, and to establish ways through which residents can draw maximum benefits from mountain tourism activities, reduce leakages associated with mountain tourism and enhance the broader circulation of mountain tourism revenue in local communities in the Eastern Highlands of Zimbabwe. The study found that local communities benefited from mountain tourism through employment opportunities, income, entrepreneurial opportunities, expanded sources of livelihood, preservation of local culture and heritage resources, clean environment, earning of foreign currency, increased taxes and levies, and support of agricultural activities necessitated by upward and backward linkages with mountain tourism in the Eastern Highlands of Zimbabwe. The findings revealed that employment was generated for residents, although it was seasonal, highly menial, and of low grades. The findings are in line with what was found by Apollo and Wengel (2022), Mbaiwa et al. (2019), and Wall and Mathieson (2006). Overall, it can be deduced that mountain tourism employment, like other subsets of tourism employment, is strongly seasonal and characterised by low-wage, part-time and casual employment, compared to many other industries. This shows that the sustainability of employment is influenced by tourism activities undertaken in the area.

More so, residents pointed out the economic benefits of mountain tourism, such as foreign currency. The residents got opportunities to sell handicrafts and showcase their culture, while the government and local authorities got taxes. The rest of the country benefits through infrastructural development, increased Gross Domestic Product, and positive word-of-mouth reviews by satisfied mountain tourists, while the tourism governing institutions such as the Zimbabwe Tourism Authority get optimum tourism levies paid by operators, and local authorities and rural district councils get various levies and taxes paid by tourism operators and other related support services. Satisfied mountain customers are prepared to pay premium prices for authentic products/services, hence increasing income for local mountain tourism service providers. On the other hand, residents reported economic changes, such as rising living costs. These findings corroborate those of Apollo and Andreychouk's (2022).

The study outlined numerous challenges that prevent local communities from benefiting fully from pro-poor mountain tourism. These challenges include the dominance of foreign companies, fragmentation of local SMEs, exclusion of local intermediaries, excessive leakages, and unsustainable jobs.

Excessive leakages were witnessed in the Eastern Highlands mountain tourism value chain, which is attributed to the dominance of foreign tourism and hospitality corporations such as tour operators, hotel chains, and travel agents, who end up siphoning revenue outside communities.

The study highlighted ways in which local communities can optimise mountain tourism benefits, such as collaboration, alliances and partnerships, and leveraging social capital. Mountain dwellers can derive maximum benefits from mountain tourism by creating strong linkages between mountain tourism activities and local community livelihoods, developing tourism packages that support and are compatible with traditional sources of livelihoods like agriculture and local culture, and involving local communities as providers of mountain tourism products as opposed to being employees. Additionally, the control of the mountain tourism activities and related tourism development processes must firmly reside in the hands of local communities, not foreign companies and other remote actors, and must give communities more voice regarding the type and nature of mountain tourism activities to be offered.

Mountain tourism needs to be aligned with communities' traditional sources of livelihoods instead of replacing them. Plugging leakages implies that income is ploughed back into the local economy. Given the multiplier effect of mountain tourism, it would create additional employment opportunities in the tourism value chain, thus stimulating demand for other services supporting mountain tourism and enhancing the circulation of money in the local economy.

Moreover, most mountain tourism resources fall under the jurisdiction of the Zimbabwe Parks and Wildlife Management Authority (ZimParks), which often disregards limits on the number of visitors to boost sales, occupancy, and capacity utilisation. The findings showed that mountain tourism and the environment conflicted. Crowded mountain spaces sometimes cause widespread environmental pollution through noise, environmental stress, breaching tolerance, and threshold capacities. Excessive noise from mountain tourists scares away wildlife, while littered plastic material endangers aquatic life and human beings as they enter the food chain and contaminate underground community water sources. These findings corroborate those of Wall and Mathieson (2006), who highlighted that tourism is highly associated with potential negative impacts on the ecosystem.

Naturally, the environment suffers, thus leading to service failures, overcrowding, staged authenticity, and dissatisfied customers. In the long haul, negative word of mouth filters through, and repeat business is not guaranteed, leading to reduced profits. Eventually, local communities become ultimate losers by suffering declining revenues and damage to their environment. Adhering to optimum visitor numbers preserves the environment and mountain tourist sites while reducing environmental stress from overcrowding. In the long run, respecting such limits is in the best interest of the community and of prospective tourists.

The Eastern Highlands have great potential to promote mountain tourism as a tool for poverty alleviation in Zimbabwe. Significant challenges prevent mountain dwellers from benefiting fully from mountain tourism; however, the challenges are not insurmountable.

Conclusion

It is evident from the research that despite the challenges being experienced by mountain dwellers, mountain tourism in the Eastern Highlands has excellent potential to alleviate poverty in line with the United Nations 2030 SDGs. It is an opportunity for the Eastern Highlands communities to benefit from their natural resource through seed money to take a leading role in mountain tourism enterprises and capacity-building programmes. Local communities should form cooperatives to run some of the hospitality and mountain tourism businesses in the area, thus reducing the dominance of foreign companies as well as reducing excessive leakages of income, if mountain tourism is to be sustainable. This might reduce the migration of local people to nearby towns such as Mutare, Harare, and Rusape. The Eastern Highlands are a unique place with a considerable potential for mountain tourism, and local communities need to engage in climate-sensitive tourism activities for long-term benefits.

Management Implications

Regardless of several challenges hindering the potential of mountain tourism in different geographical contexts, there is a need for capacity development, especially among local communities, on how best to boost mountain tourism. There is a need to engage in climate-sensitive tourism activities that are strongly linked to the local economy to reduce leakages if mountain tourism is to be sustainable. Suppose mountain tourism is to be a vehicle for poverty eradication in mountain communities. In that case, all stakeholders need to collaborate and focus on the present and future social, environmental, and economic developmental effects of mountain tourism to address the needs of residents, service providers, and the government.

The packaging of mountain tourism product offerings must complement the local economy for maximum benefits. Similarly, tourism activities must complement and support the traditional sources of livelihood and have strong linkages with the local economy by supporting entrepreneurial activities by local communities.

Locally owned SMEs must control the mountain tourism value chain to reduce revenue leakage. To enhance benefits, coommunities need to collaborate instead of competing. Also important is environmental protection, which ensures that tourist numbers do not exceed the required threshold. In the same vein, preserving culture, heritage resources, and the environment guarantees long-term benefits for local current and future generations.

References

Apaza-Panca, C.M., Arevao, J.E.S, C. and Apaza-Apaza, S. (2020). Agritourism: alternative for sustainable rural development. *Domino de las Ciencias* 6(5), 207–227. https://doi.org/10.23857/dcv6i5.1596.

Apollo, M. (2021). *Environmental impacts of mountaineering: A conceptual framework.* Springer. https://doi.org/10.1007/978-3-030-72667-6

Apollo, M., & Andreychouk, V. (2022). *Mountaineering Adventure tourism and local communities: Social, environmental and economic interactions.* Edward Elgar Publishing. https://doi.org/10.4337/9781802209389.

Apollo, M., & Wengel, Y. (2022). Mountaineering tourism. In D. Buhalis (Ed.), *Encyclopaedia of tourism management and marketing*. Edward Elgar Publishing, DOI:10.4337/9781800377486.mountaineering.tourism.

Beedie, P., & Hudson, S. (2003). Emergence of mountain-based adventure tourism. *Annals of Tourism Research*, 30(3), 625–643. https://doi.org/10.1016/S0160-7383(03)00043–00044.

Borinodo, T., & Coromina, L. (2018). Social capital as causes and consequence of rural touristic development: The case of Lindogo (Portugal). *European Journal of Tourism Research* 19, 71–85.

Carney, D. (1999). *Approaches to sustainable rural livelihoods for the rural poor.* Overseas Development Institute.

Chibaya, T., & Zhou, Z. (2023). Impact, recovery and resilience of the events and festivals sector from COVID-19 crisis: lessons from Zimbabwe. In Dube, K., Mensah, I., Chapungu L. (eds) *COVID-19, Tourist Destinations and Prospects for Recovery.* Springer. https://doi.org/10.1007/978-3-031-24655-5_7.

Chipfuva, T. (2021) *Developing a pro-poor tourism framework for Zimbabwe* (PhD Thesis North West University, South Africa).

Gohori, O., & van der Merwe, P. (2022b). Limitations to community participation in tourism from local people's perspectives: Manicaland province, Zimbabwe. *Development Southern Africa* 39(6), 841–855.

Gohori, O., & van der Merwe, P. (2020). Towards a tourism and community-development framework: An African perspective. *Sustainability* 12(13), 5305. https://doi.org/10.3390/su12135305.

Häuberer, J. (2011). *Social Capital Theory.* VS Verlag für Sozialwissenschaften.

Kinsey, H. (2010). *Poverty dynamics in rural Zimbabwe: The 30 years (lost). War against poverty.* Chronic Poverty Research Centre, University of Manchester.

Lak, A., & Khairabadi, O. (2022) Leveraging agritourism in rural areas in developing countries: The case of Iran. *Frontiers on Sustainable Cities*, 4, 863385. doi:10.3389/frsc.2022.863385

Logan, B. I., & Mosely, W.G. (2002). The ecology of poverty alleviation in Zimbabwe's campfire. *Geoforum* 33(1), 1–14.

Lurdi, E., & Bird, K. (2007). *Understanding poverty.* Retrieved from https://www.poverty-wellbeing.net.

Makoni, L.P. (2014). From financial exclusion to financial inclusion through micro finance: The case of rural Zimbabwe. *Corporate Ownership and Control* 11(4), 447–415.

Manjengwa, J., Matema, C., & Tirivanhu, D. (2016). Understanding urban poverty in two high density suburbs of Harare, Zimbabwe. *Development Southern Africa* 33(1), 23–38. doi:10.1080/0376835x.2015.1116376.

Mbaiwa, J. E., Mbaiwa, T., & Siphambe, G. (2019). The community-based natural resource management programme in southern Africa–promise or peril? The case of Botswana. In M. Mkono (Ed.), *Positive Tourism in Africa* (11–22). Routledge.

Moscardo, G. (2015). Social capital trust and tourism development: In R. Nunkoo & S.C.J. (Eds.), *Trust, Tourism Development and Planning (66–68*). Routledge.

Moscardo, G., Kanovalova, E., Murphy, L., McGeheeb, N.G., & Scharmanna, A. (2017). Linking tourism to social capital in destination communities. *Journal of Destination Marketing and Management* 6, 286–295. https://doi.org/10.1016/j.jdmm.2017.10.001.

Musasa, G., & Mago, S. (2014). Challenges of rural tourism development in Zimbabwe: a case of the Great Zimbabwe-Masvingo district. *African Journal of Hospitality, Tourism and Leisure*, 3(2), 1–12.

Musavengane, R. (2018). Toward pro-poor local economic development in Zimbabwe: the role of pro-poor tourism. *African Journal of Hospitality, Tourism and Leisure*, 7(1), 1–15.

Mutana, S., Chipfuva, T., & Muchenje, B. (2013). Is tourism in Zimbabwe developing with the poor in mind? Assessing the pro-poor involvement of tourism operators located near rural areas in Zimbabwe. *Asia Social Science* 9(5), 1–9.

Price, M., Wachs, T., & Byers, E. (1999). *Mountains of the world: Tourism and sustainable mountain development*. Institute of Geography, University of Berne.

Pro-Poor Partnership, (2004). *Tourism and poverty reduction – Making the links*. https://silo.tips/downloads/pro-poor-tourism-info-sheets.

Rogerson, C.M. (2006). Pro-poor local economic development in South Africa: The role of pro-poor tourism. *Local Environment* 11(1), 37–60.

Scheyvens, R. (Ed.) (2014). *Development fieldwork: a practical guide*. Sage. https://doi.org/10.4135/9781473921801.

Scheyvens, R. (2007). Exploring the tourism-poverty nexus. *Current Issues in Tourism* 10(2/3), 231–254. https://doi.org/10.2167/cit318.0.

Spenceley, A., & Meyer, D. (2012). Tourism and poverty reduction: Theory and practice in less economically developed countries. *Journal of Sustainable Tourism* 20(3), 297–317. https://doi.org/10.1080/09669582.2012.668909.

UN, (2017). *Sustainable Development Goal 8*. Accessed 06/05/2023. https://sustainabledevelopment.un.org/sdg8.

UNWTO, (2018). *UNWTO Tourism highlights*, 2018 edition.

Wall, G., & Mathieson, A. (2006). *Tourism: Change, impacts and opportunities*. Pearson.

Walters, T. (2016). Using Thematic Analysis in Tourism Research. *Tourism Analysis*, 21(1), 107–116. https://doi.org/10.3727/108354216X14537459509017.

Zhou, Z. (2018). The tourism sector: A bright light in Zimbabwe's depressed economic environment *African Journal of Hospitality, Tourism and Leisure*, 7(1): 1–19.

Zhou, Z., & Chibaya, T. (2022) Towards Gender and Disability Inclusive Tourism Development in Zimbabwe: Insights through sustainable development goals. In: Woyo, E., & Venganai, H. (Eds.), *Gender, Diversity Disability, and Tourism in Africa*. Macmillan. https://doi.org/10.1007/978-3-031-12551-5_7.

11 Ethnic Tourism

Challenges and Opportunities for Poverty Alleviation Based on Highlands of Lao Cai, Northwest Vietnam

Phuong Bui L.A and The-Bao Luong

Introduction

According to Goodwin (2009), the shift to a free-market system, which occurred years after Thatcherism, has resulted in the rise of pro-poor tourism. The first Sustainable Development Goal (SDG) is to eradicate poverty in all forms, and tourism is recognised as a potential means to achieve this (UNWTO – Tourism for SDGs). However, the implementation of tourism for poverty alleviation faces challenges such as increasing income disparities among tourism workers, limited training and career opportunities, and average salaries with non-monetary benefits (SDGs 8 and 10) (Kronenberg & Fuchs, 2021). Despite these challenges, tourism holds the potential to be an effective tool in achieving UN SDG 1 and combating poverty.

This chapter provides fresh insights into tourism development in Sapa, Northern Vietnam, a mountainous region with natural landscapes, vibrant cultural diversity, and challenging conditions (Apollo & Wengel, 2022). It explores how pro-poor tourism is applied in this district, utilises the tourism global value chain (GVC) to investigate the engagement of locals in the production of tourism products, as well as the relationship between pro-poor tourism inside the GVC, which has not been done in many critical tourism studies. The GVC model, originating in microeconomics, analyses how firms source resources from various regions to produce their products or services (Song et al., 2012) and involves fieldwork and in-depth interviews to understand the previously described relationship better.

Background

Pro-poor tourism is one of the tourism approaches that aims to provide a wide range of benefits to people experiencing poverty, including social, environmental, and cultural benefits to developing countries (Mitchell, 2012). The primary goal of pro-poor tourism is to increase impoverished people's access to economic advantages, eliminate negative social and environmental repercussions, and implement policy and procedural changes (Roe & Urquhart, 2001). By offering chances, pro-poor tourism attempts to provide net benefits for the

DOI: 10.4324/9781003392583-13

poor within the tourism industry. It is not limited to a specific product or sector but rather a strategy to address poverty (Sudsawasd et al., 2022). It aims to promote tourism development, help stabilise livelihoods and alleviate poverty in impoverished areas (Folarin et al., 2020; Sharma et al., 2021).

Pro-poor tourism puts poor people and poverty at the centre of the sustainability debate to examine how tourism generates net benefits (Buhalis et al. 2023). In addition, It is difficult to accurately assess its impact on national poverty alleviation efforts (Oviedo-García et al., 2018).

Wen et al. (2021) suggest that pro-poor tourism has a dual influence on various stakeholders:

- it affects local government by fostering social development and providing policy support,
- it impacts tourism enterprises by facilitating capital input ensuring profit return, and
- it influences community residents by encouraging community participation and fostering community development.

According to Ashley & Roe (2001), the government can consult communities, stimulate private sectors, adjust rules, integrate plans, promote tourism, and remove impediments to establishing poverty-alleviating tourism. Similarly, Ji et al. (2022) concluded that the government plays a significant role in establishing links that promote local participation and income in the tourism business, thus lowering poverty.

Tourism firms play an important role in pro-poor tourism by leveraging resources from rural areas with high poverty levels and implementing social responsibility activities, consequently creating economic opportunities (Medina-Muñoz et al., 2016). However, challenges remain regarding improving local procurement, labour conditions, and equitable income distribution. Current tourism policies favour large foreign-owned resorts, limiting support for indigenous businesses and hindering the sector's pro-poor potential (Scheyvens & Russell, 2012). Disparities in benefits have also emerged, separating communities into tourist and non-tourist zones (Nguyen & Funck, 2019).

Pro-poor tourism aims to generate economic benefits for the disadvantaged and create sustainable and inclusive tourism actions that help develop and empower local communities (Okazaki, 2008). This endeavour intersects with the global value chain (GVC) approach, which has been applied to examine the tourism industry in various places worldwide (Bire & Nugraha, 2022; Christian & Nathan, 2013; Staritz & Reis, 2013). By utilising the GVC model, researchers can delve into how locals engage in the production of tourism products within these communities and explore the intricate relationship between pro-poor tourism initiatives and their integration within the broader GVC framework.

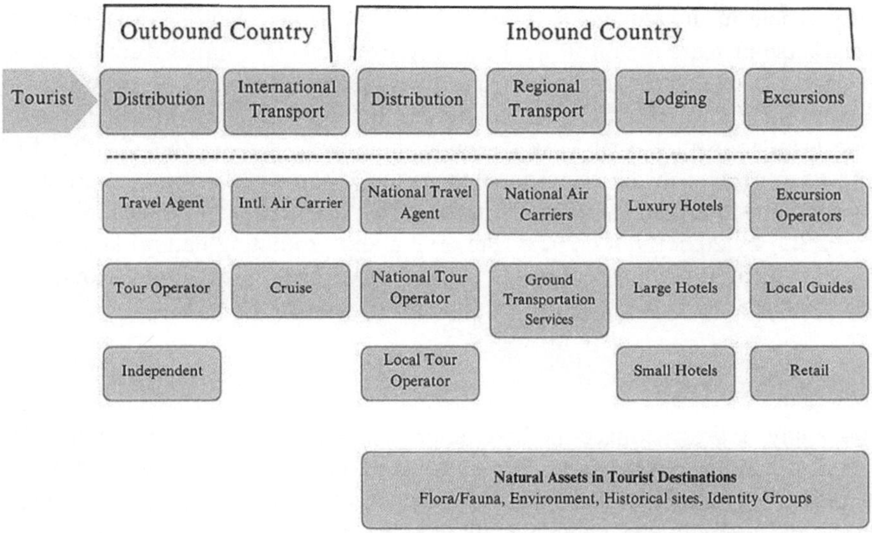

Figure 11.1 Global Value Chain (GVC) in tourism
Source: Redrawn from Bire & Nugraha (2022); Christian & Nathan (2013); Staritz & Reis (2013)

Study Area

Sapa, nestled in the highlands of the Sapa district within Lao Cai province, Vietnam, spans a natural expanse of 68,329 hectares. Covered in clouds year-round, Sapa's average altitude ranges from 1,400 to 1,700 meters above sea level. The terrain features dissected landscapes, steep slopes, and narrow valleys, making it inherently hazardous.

Sapa is home to various ethnic minorities, including H'Mong-54.9%, Giao-25.6%, Kinh-13.6%, Tay-3%, Giay-1.6% and others (Quốc Hồng, 2022). With the H'Mong being the predominant community, traditionally dependent on agriculture and forestry, the H'Mong have expanded into tourism and related sectors while maintaining their traditional farming and craftsmanship practices (Thanh-Tuan, 2021).

Tourism in Sapa is heavily affected by the Kinh, who profit from the increase of tourism, and external investors like the Sun Group, which are the powerful elite who hold sway over tourism development in the region, likely leading to diverging perspectives on this growth (Michaud & Turner, 2006).

Support for tourism in Sapa is multifaceted, drawing from a spectrum of stakeholder motivations. While economic considerations are prominent among Kinh stakeholders, non-economic factors also weigh significantly, notably for ethnic minorities who emphasise sociocultural elements such as cultural preservation, recognition, self-identity, and community solidarity (Boley & McGehee, 2014). Nonetheless, despite the diversity of motivations, economic advantages emerge as the predominant driving force behind the enthusiastic endorsement of tourism development in the region (Nguyen et al., 2023). Sapa Town has experienced a remarkable increase in visitor numbers, with over 534,000 tourists recorded since the start of 2023, triple the figure from the same period in 2022, as reported by Hoang Thi Vuong, Head of the Culture and Information Department. The revenue from tourism has also seen significant growth, surpassing 3.2 times the amount generated in the corresponding period last year. The local government expects tourism revenue in Sapa Town to reach 13,000 billion VND (ca. 530 million USD) by 2023 (NDO, 2023). Preserving and promoting natural landscapes and architectural landmarks like the Sapa terraced fields, Ham Rong Mountain, and the stone church are also key focal points of the plan (VNMC).

Methodology

This study utilises a mixed-methods approach, including secondary materials, interviews, and fieldwork research to understand, observe, and interact with Sapa people in their natural setting. The 46 participants (Tab. 11.1), comprising 32 ethnic people and 14 Kinh-lowland individuals working in various roles, were interviewed to understand their roles, opportunities, and challenges in Sapa tourism.

Interviews with individuals were conducted to understand distinctions in the roles of Kinh people and ethnic minorities in the GVC for tourism in Sapa. Following Wen's et al. (2021) pro-poor tourism model, interviews focused on the following:

- Community residents and pro-poor tourism,
- Local government and pro-poor tourism, and
- Tourism enterprise and pro-poor tourism.

Fieldwork was conducted at popular tourist sites, and qualitative analysis techniques, such as content and thematic analysis, were employed. Findings were compared with existing literature on sustainable tourism development.

Results

Through interviews and examination of interviewees' documents, this study delved into the establishment of pro-poor tourism in the Sapa mountain

Table 11.1 Interviewee Features

No	*Interviewee categories*	*Title*	*Interviewee Features*	*No of interviewee*	*Age*
1	ID 1 – ID 32	Ethnic people Giao Do (ID1–11), H'mong (ID12–32)	Rice farmers, tour guides, peddlers	32	15–40
2	ID 33 – ID 46	Kinh people	Employees/ Employers	14	25–42

region. Two overarching themes emerged: Sapa community participation and Sapa community development.

Sapa community participation

Ethnic people's participation rates are relatively low, and the community is introduced to work opportunities through initiatives run by the local government. For example, ID 1 said: "Some of us are introduced to finding jobs by the local government, participating in training programs with prominent entities like Sun Group" (ID numbers 2,3,4,5,11,12,13,14, and 22.)

Some young H'mong with efficient English work as tour guides; for example, ID 12 said (similar response was given by research participants ID numbers 13-32):

> "Our H'Mong members actively participate in guiding work, serving as freelance guides. Our responsibility involves leading guests on mountain expeditions. However, it is important to note that young individuals with fluent English skills typically take on this role, as it requires dedicating several days to travel and having basic first-aid abilities."

Women from the Giao Do minority work in tourism as employees and handicraft street hawkers. For example, ID2 shared that "I'm Giao Do, I choose selling brocade products to tourists as a peddler, my children follow me to sell them too, and my husband often focuses on tending to livestock in the nearby mountains." (ID numbers 1,3 and 6)

Individuals from lowland places who are well-educated, such as Hai Phong and Gia Lai, have moved to Sapa and acquired specialised jobs in industries such as hospitality. Some have even launched their enterprises, which help to boost the local economy and provide jobs. Members of the Kinh ethnic group, originating from Hai Phong and Gia Lai, emphasise the role of education in providing enhanced employment prospects compared to ethnic minority groups. After graduating from universities, they relocated to Sapa, where they secured skilled positions within the hospitality sector, primarily in hotels and restaurants. Additionally, some individuals pursued entrepreneurial endeavours, establishing various start-up businesses, including restaurants, small-scale or boutique hotels, and spa facilities, assuming ownership of these establishments.

Ethnic individuals find employment in lower-skilled positions as housekeepers or spa therapists. This diversity reflects the different opportunities available for ethnic people within the local area.

Some participants articulated their viewpoints concerning education and employment opportunities, highlighting the barriers faced by women in their community. They noted that women often confront difficulties accessing formal education and training, which restricts them to roles primarily as housekeepers in hotels, spas, or other service-oriented establishments. The attainment of positions in skilled roles such as receptionist or supervisor remains elusive due to these limitations. Furthermore, other participants (ID 22–32) echoed comparable views during the interview sessions. They highlighted that the limited access to education confines some community members to domestic duties, primarily focusing on farming and household management. These circumstances exacerbate economic challenges, contributing to the perpetuation of poverty within their community.

Aside from traditional salaried employment, certain ethnic minorities have ventured into the homestay business model despite financial constraints. They usually offer complimentary accommodation to guests who opt to hire them as tour guides, whereas those who do not incur an accommodation fee of approximately 20 USD per night. Furthermore, challenges are evident concerning investment and the maintenance of quality standards. For instance, participants with ID 24 and 29 revealed that ethnic individuals in Sapa face difficulties in establishing their businesses, particularly in offering homestay services, due to financial constraints. However, despite their efforts, limited investment and insufficient attention to essential conditions and hygiene standards occasionally lead to these homestays failing to provide satisfactory accommodations.

Sapa community development

Sapa community development encompasses three primary domains: economic, social, and environmental. The tourism industry in Sapa has both benefits and challenges to the local community development. Participants mentioned that many residents have observed firsthand the significant impact of tourism employment and service income on improving numerous families' material and spiritual well-being. There is a diversity of Sapa homestays owned by Kinh people, foreigners, and locals, each varying in investment and management approaches. Kinh people and foreigners typically possess better education and resources and interact with tour companies online. In contrast, indigenous residents tend to depend more on local guides and agencies, especially for short-stay tourists.

Tourism significantly impacts Sapa's social development, revitalising traditional handicrafts. Furthermore, participants mentioned that traditional festivals like the "Tết Nhảy" of the Red Dao were revitalised due to tourism. Giay people mentioned that "Lễ hội xuống đồng" (going down to the fields ceremony), along with similar festivals of other ethnic groups in Sapa, has experienced a revitalisation through tourism. As tourism has grown in the region, there has been increased interest from visitors in experiencing and participating in authentic cultural events and traditions. Consequently, local communities, including the Giay people, have often integrated elements of their traditional festivals into tourism activities, allowing visitors to witness and engage with these cultural celebrations. This integration does not only help to preserve and promote the cultural heritage of the Giay ethnic group but also provides economic opportunities for the community through tourism-related activities.

Furthermore, it has been observed that influences from Western and Hanoi cultures consequently reshaped local traditions. Notably, increased tourist traffic on Sapa's central street creates livelihood opportunities for selling handmade products near the Stone Church, sometimes using imported materials to meet the large number of tourists' demand (ID numbers 13,14,15,16.)

Tourism developments in Sapa have been impacted by the sustainable tourism concept, leading to positive changes like using gas or electricity for cooking and heating. However, challenges persist in waste management and wastewater treatment. Participants from the hospitality sector noted that despite some recycling efforts, non-biodegradable waste accumulates due to the high number of tourists, which harms the environment. ID 33–46 shared, "living in Sapa, we face challenges with inadequate wastewater treatment, leading to septic tank waste entering water bodies and small streams."

Pro-Poor Tourism in Sapa and Local Government

In Lao Cai province, Sapa boasts an 85% ethnic minority population. Despite that, there is limited participation in tourism plans. Local Kinh individuals have highlighted that language differences and a lack of understanding of tourism processes can marginalise ethnic communities. However, some officials highlight the government's support for cultural activities and festivals and offer residents loans. Despite initial hesitation, some locals, particularly women, have embraced entrepreneurship, leveraging their exposure to tourism and strong language skills. Furthermore, other women collaborate with the Commune Women's Union to develop tourism services and engage in training activities to support their businesses and align with local needs and sustainable trends (ID numbers 6,8,11).

Sapa tourism enterprise

Tourism growth in Sapa drives up real estate prices, attracting profits for early investors. With its unique climate and landscape, the rise of homestays and resorts in this mountainous region results in high tourism demand. Sapa is considered an untapped "gold mine," presenting significant investment opportunities in mountainous real estate development. According to local authorities and statistics, Sapa currently has 571 accommodation establishments and 6,800 rooms, along with 400 restaurants and various services. However, there is a lack of centralised shopping and entertainment areas, and accommodation facilities often face overcrowding issues, especially during vacations and holidays.

Despite the challenges mentioned in this chapter, several young ethnic individuals have taken the initiative to establish their businesses. For instance, one participant expressed their desire for support from the local government to establish a personal company after studying in the delta region. Another participant noted that many married women are actively involved in developing enterprises focused on indoor agricultural products, such as orange gardens, which appeal to clients seeking unique experiences and deeper insights into indigenous cultures. Furthermore, a group of participants highlighted that H'Mong women are investing in opening homestays and selling brocades and souvenirs. They mentioned that during weekends and Tet holidays when there is an influx of tourists, these women take a break from their agricultural work to welcome visitors. As a result, the economic status of women has improved, leading to a significant enhancement in the position of H'Mong women within their families and society compared to previous circumstances.

The interviews underscore the proactive initiatives of the Sapa government aimed at fostering youth economic development. Programs such as the "Accompanying Young People in Start-up Activities" movement and vocational training initiatives have been implemented to facilitate job placements and foster innovative start-up models, including community tourism projects and homestay concepts centred around fruit gardens. These efforts demonstrate the government's dedication to empowering young individuals and fostering sustainable economic advancement in the region. One participant shared: "Sapa's Youth Unions and Associations have successfully executed the "Accompanying Young People in Start-up Activities" programme, actively cooperating with relevant departments to organise employment fairs and advising the town People's Committee on economic development projects. Participants highlighted their involvement in vocational training programs aimed at young people in ethnic minority areas, which involve partnerships with specialised departments for consultation and career orientation from Youth Unions.

Additionally, other participants shared observations of various community tourism models. For instance, one model involves a 3.4 km check-in point project within a 3-hectare forest, providing employment opportunities for 20 young individuals. Another successful model integrates homestays in Tay stilt houses with fruit gardens and the production of tourism goods such as brocade, teas, and herbs.

The Global Value Chain for Tourism in Sapa

The interviews highlight clear distinctions in the roles of Kinh people and ethnic minorities in Sapa's tourism. Despite making up 85% of the population, ethnic minorities primarily work as hired labourers or run small businesses like homestays and serve as local freelance guides with fluctuating incomes, catering to inbound tourism. Meanwhile, the major investments come from lowland – Kinh people, driven by big enterprise groups and individuals from the delta area (see Figure 11.2).

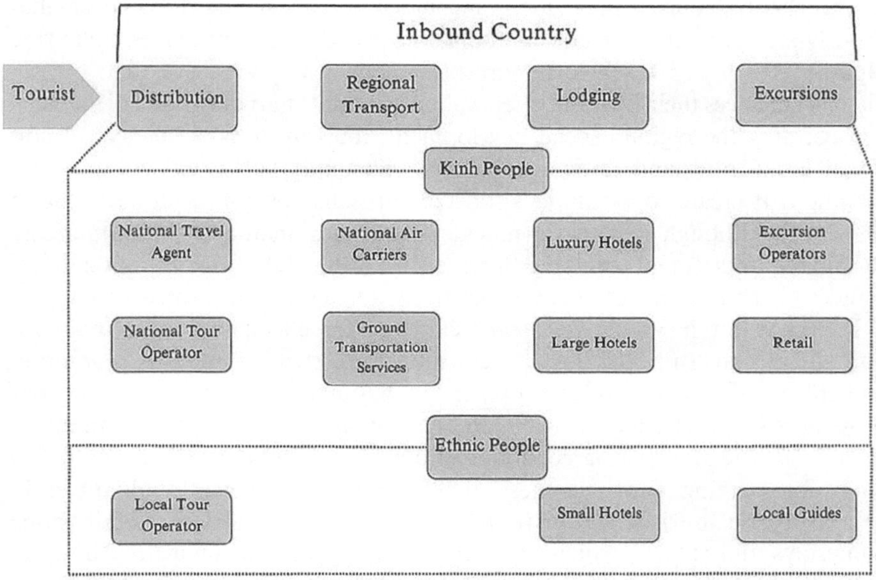

Figure 11.2 Global Value Chain (GVC) for tourism in Sapa, analysis by author, 2023

Discussion

Using Wen et al.'s (2021) multidimensional pro-poor tourism theory, this research examines the viewpoints of ethnic and lowland residents in Sapa, a highland region in northern Vietnam, on the impact of tourism on poverty reduction. The objectives of this article are twofold:

1 to investigate variations in the perceptions of ethnic groups and lowland people regarding the three dimensions of pro-poor tourism and global value chains (GVC) and
2 to develop a GVC model specific to pro-poor tourism in Sapa to explore the distribution of employment opportunities as a means of poverty elimination, addressing both the opportunities and challenges.

Opportunities

Community participation and development have the potential to contribute to Sapa's tourism development. Tourism activities in Sapa are geared towards benefiting the local community by providing economic opportunities to alleviate poverty (Jönsson, 2012; Spenceley, 2022) and improving livelihoods (Truong et al., 2014). This is achieved through initiatives such as offering homestay accommodations, providing local guiding services, and showcasing cultural traditions and handicrafts (Truong et al., 2014). These endeavours aim to elevate the local population's income and living standards while simultaneously preserving their unique cultural heritage and natural resources. Following Nguyen et al. (2023), ethnic minorities in Sapa strongly believe that tourism directly enhances their lives and offers valuable insights into the [tourism] industry.

Regarding the region's social development, tourism in Sapa enhances traditional handicrafts and festivals, providing economic value and cultural preservation. It creates opportunities for local artisans, fostering cultural pride in the world. Although some environmental challenges (mainly pollution) remain, sustainable tourism practices in Sapa, such as the shift from traditional fuel sources to eco-friendly alternatives, positively impact the local environment.

In this research, we provide some evidence for effective policies that support the region. In Sapa, government initiatives, such as cultural promotion and interest-free loans, offer support, but hesitancy due to repayment concerns persists among locals. The Commune Women's Union acts as an entrepreneurial bridge, fostering economic and women's empowerment, and it has plans for training and awareness aligned with sustainable trends. Overall, Sapa's diverse tourism landscape has spurred unique Sapa mountain slope homestays and resorts, which are seen as untapped gold mines for entrepreneurial initiatives, particularly among ethnic communities. Such tourism development has led to innovative start-ups, empowered women economically, and fostered youth economic development.

Challenges

Concerning community participation and development, some challenges remain, especially for ethnic women, facing barriers in education and training access, limiting them to traditional roles (Freund & Hernandez-Maskivker, 2021) that, coupled with economic hardships, perpetuates a poverty cycle in some communities and women empowerment. Kinh stakeholders place greater importance on tourism for local economic growth than ethnic minorities (Nguyen et al., 2023). Tourism planning responsibility rests with local authorities and corporations like Sun Group (Nguyen et al., 2020). Village representatives merely implement received instructions without contributing to the planning process. The growing job opportunity gap between the Kinh-lowland people and ethnic communities underscores the need for inclusive policies that significantly impact poverty alleviation (Quyen & Tuan, 2022).

Local government tends to favour non-local investors, leaving limited returns for the local population (Wu et al., 2011). Ventures into homestay investments by ethnic individuals face hurdles due to financial constraints and hygiene standards, demanding focused support in acquiring knowledge and loan capacity (Truong et al., 2014). Tourism technology use varies among Kinh people, foreigners, and ethnic groups. Kinh and foreigners excel in online interactions with travel companies, while ethnic minorities face technological limitations, relying more on guides. This discrepancy can worsen economic disparities, necessitating a comprehensive approach for equitable community benefits.

During the fieldwork and interviews, it was apparent that Sapa's tourism relies on imported souvenirs for traditional handicrafts, often sold at higher prices by some ethnic individuals who misrepresent them as local products. This practice, driven by adapting to tourist demands, commodifies and distorts local cultures and destinations (Trupp et al., 2023).

The influx of investment and job-seeking individuals from the delta, including Kinh people and foreigners, has led to the erosion of local culture (Yang et al., 2023). For instance, the influence of Western and lowland culture (for example, from Hanoi) poses a challenge to the authenticity of local traditions. Local government must navigate this challenge by supporting and encouraging collaboration to preserve authentic festivals, balancing economic gains and cultural issues.

Tourism also brought some environmental impacts. The influx of tourists has created environmental challenges, including waste management issues and pollution from inadequate wastewater treatment. In reality, aligning tourism with environmental goals and minimising negative social impacts is feasible with fewer tourists, albeit with lower community income (Apollo et al., 2022). Efforts are needed to address waste and sanitation issues for sustainable tourism in Sapa (Yang et al., 2021).

Challenges related to policy support highlight the need for inclusive decision-making processes, addressing resource allocation inequities, and combining economic opportunity with cultural preservation. Bridging talent and technology gaps, particularly among young entrepreneurs, is critical to the success of pro-poor tourism in Sapa.

Among ethnic minorities, both government-supported and self-initiated businesses are trends in common. However, giving critical training and skills to young and female entrepreneurs continues to be a significant difficulty. These include restricted access to resources and educational opportunities, cultural hurdles that may limit women's engagement in entrepreneurship, and the necessity for bespoke training programmes that address the specific requirements and limits experienced by these groups. Furthermore, geographical isolation and a lack of infrastructure might impede efforts to reach and encourage budding entrepreneurs, particularly in rural areas. Overcoming these obstacles necessitates focused interventions such as mentorship programmes, financial access, and activities that promote gender equality and empower marginalised populations.

Conclusion

Sapa's booming tourism industry presents a complex landscape of both opportunities and challenges in terms of poverty reduction. Tourism may produce revenue and job possibilities for community residents by highlighting the distinct cultural traditions, handicrafts, and lifestyles of several Sapa ethnic groups. Furthermore, developing and marketing ethnic tourism programmes can instil a sense of pride and ownership in local communities, motivating people to participate actively in and benefit from the tourism industry. While it holds great promise for economic development among ethnic minorities, it also faces the difficulty of conserving cultural history, addressing skill gaps, and ensuring environmental sustainability.

Management Implications

In Sapa, as in many other parts of the world, enacting inclusive policies that balance economic development and cultural preservation is vital to the long-term viability of pro-poor tourism programmes. Targeted training and capacity-building programmes can provide individuals with the skills and information they need to participate and benefit from the tourism business effectively. This covers instruction in hospitality, customer service, language skills, and sustainable tourism practices. Area management should encourage tourists to practise responsible tourism behaviours such as respecting local customs and traditions, supporting locally owned companies, and reducing negative environmental consequences, they should prioritise conservation efforts such as sustainable land management methods, waste management initiatives, and biodiversity protection strategies

Acknowledgement: Thanks to the SILKEN Vietnam Project, funded by the British Council with Sapa research initiatives.

References

Apollo, M., Andreychouk, V., Rawat, K., Mostowska, J., Jones, T. E., Rettinger, R., & Maciuk, K. (2022). Himalayan nature-based tourism. Challenges for tourism and protected areas. *International Journal of Conservation Science*, 13(1), 249–266.

Apollo, M., & Wengel, Y. (2022). *Mountaineering Tourism: A Critical Perspective.* Routledge. https://doi.org/10.4324/9781003095323.

Ashley, C., & Roe, D. (2001). *Pro-poor tourism strategies: Making tourism work for the poor: A review of experience.* Overseas Development Institute.

Bire, R. B., & Nugraha, Y. E. (2022). A value chain perspective of the new normal travel behaviour: A case study of Indonesian millennials. *Tourism and Hospitality Research*, *22*(4), 462–472. https://doi.org/10.1177/14673584211065615.

Boley, B. B., & McGehee, N. G. (2014). Measuring empowerment: Developing and validating the Resident Empowerment through Tourism Scale (RETS). *Tourism Management*, 45, 85–94. https://doi.org/10.1016/J.TOURMAN.2014.04.003.

Christian, M. M., & Nathan, D. (2013). Tourism Overview: Changing End Markets and Hyper Competition. *SSRN Electronic Journal.* https://doi.org/10.2139/SSRN.2237497.

Folarin, O., Research, O. A.-J. of T., & 2020, undefined. (2020). Does tourism reduce poverty in Sub-Saharan African countries? *Journal of Travel Research* , 59(1), 140–155. https://doi.org/10.1177/0047287518821736.

Freund, D., & Hernandez-Maskivker, G. (2021). Women managers in tourism: Associations for building a sustainable world. *Tourism Management Perspectives*, 38, 100820. https://doi.org/10.1016/J.TMP.2021.100820.

Goodwin, H. (2009). Reflections on 10 years of Pro-Poor Tourism., *Journal of Policy Research in Tourism, Leisure and Events*, 1(1), 90–94. http://Dx.Doi.Org/10.1080/19407960802703565.

Ji, Z., Xu, H., & Cui, Q. (2022). Tourism and poverty alleviation in Tibet, China: the role of government in enhancing local linkages. *Asia Pacific Journal of Tourism Research, 27*(2), 173–191. https://doi.org/10.1080/10941665.2021.2020312.

Jönsson, C. (2012). Understanding the Dimensions of Pro-Poor Tourism. *Tourism Geographies*, 14(2), 356–358. https://doi.org/10.1080/14616688.2012.633221.

Kronenberg, K., & Fuchs, M. (2021). Aligning tourism's socio-economic impact with the United Nations' sustainable development goals. *Tourism Management Perspectives*, 39, 100831. https://doi.org/10.1016/J.TMP.2021.100831.

Medina-Muñoz, D. R., Medina-Muñoz, R. D., & Gutiérrez-Pérez, F. J. (2016). A Sustainable Development Approach to Assessing the Engagement of Tourism Enterprises in Poverty Alleviation. *Sustainable Development*, 24(4), 220–236. https://doi.org/10.1002/SD.1624.

Michaud, J., & Turner, S. (2006). Contending visions of a hill-station in Vietnam. *Annals of Tourism Research*, 33(3), 785–808. https://doi.org/10.1016/J.ANNALS.2006.02.003.

Mitchell, J. (2012). Value chain approaches to assessing the impact of tourism on low-income households in developing countries. *Journal of Sustainable Tourism*, 20(3), 457–475. https://Doi.Org/10.1080/09669582.2012.663378.

Nguyen, H. Van, Diane, L., & Newsome, D. (2020). Kinh and ethnic tourism stakeholder participation and collaboration in tourism planning in Sapa, Vietnam. *International Journal of Culture, Tourism, and Hospitality Research*, 14(4). https://doi.org/10.1108/IJCTHR-12-2018-0179.

Nguyen, H. Van, Lee, D., & Warren, C. (2023). A comparison of stakeholder perspectives of tourism development in Sapa, Vietnam. *Tourism and Hospitality Research*, 23(1), 17–29. https://doi.org/10.1177/14673584221075179.

Nguyen, H. Van., & Funck, C. (2019). Tourism's Contribution to an Equal Income Distribution: Perspectives from Local Enterprises. *Tourism Planning & Development*, 16(6), 637–656. https://Doi.Org/10.1080/21568316.2018.1563564.

Okazaki, E. (2008). A community-based tourism model: Its conception and use. *Journal of Sustainable Tourism*, 16(5), 511–529. https://doi.org/10.1080/09669580802159594.

Oviedo-García, M. Á., González-Rodríguez, M. R., & Vega-Vázquez, M. (2018). Does Sun-and-Sea All-Inclusive Tourism Contribute to Poverty Alleviation and/or Income Inequality Reduction? The Case of the Dominican Republic. *Journal of Travel Research*, 58(6), 995-1013. https://doi.org/10.1177/0047287518789272.

Quoc, Hong (2022). *Phát triển du lịch cộng đồng ở Sa Pa*. The official website of the Central Committee of the Communist Party of Vietnam. Retrieved from: https://nhandan.vn/phat-trien-du-lich-cong-dong-o-sa-pa-post723360.html.

Roe, D., & Urquhart, K. (2001). *Pro-Poor Tourism: Harnessing the World's Largest Industry for the World's Poor*. World Summit on Sustainable Tourism.

Scheyvens, R., & Russell, M. (2012). Tourism and poverty alleviation in Fiji: comparing the impacts of small- and large-scale tourism enterprises. *Journal of Sustainable Tourism*, 20(3), 417–436. Https://Doi.Org/10.1080/09669582.2011.629049.

Sharma, M., Mohapatra, G., & Giri, A. K. (2021). Is tourism pro-poor in India? An empirical investigation using ARDL approach. *Journal of Economic and Administrative Sciences*, 38(4), 602–618. https://doi.org/10.1108/JEAS-02-2021-0031.

Song, H., Liu, J., & Chen, G. (2012). Tourism Value Chain Governance: Review and Prospects. *Journal of Travel Research*, 52(1), 15-28 http://Dx.Doi.Org/10.1177/0047287512457264.

Spenceley, A. (2022). Pro-Poor Tourism's Evolution and Implications Arising from the COVID-19 Pandemic. *Tourism Planning & Development*, 19(1), 13–25. https://doi.org/10.1080/21568316.2021.2021470.

Staritz, C., & Reis, J. (2013). *Global value chains, economic upgrading, and gender: Case studies of the horticulture, tourism, and call center industries.* The World Bank. http://hdl.handle.net/10986/16976.

Sudsawasd, S., Charoensedtasin, T., Laksanapanyakul, N., & Pholphirul, P. (2022). Pro-poor tourism and income distribution in the second-tier provinces in Thailand. *Area Development and Policy*, 7(4), 404–426. https://Doi.Org/10.1080/23792949.2022.2032227.

Thanh-Tuan. (2021). *Portrait of 05 ethnic groups in Sapa*. Sapa Tourism Office. https://sapa-tourism.com/first-time-sapa/sapa-the-best-choice-for-holiday-on-summer/.

Quyen, L. T. T., & Tuan, T. H. (2022). Tourism impacts of poverty alleviation on ethnic households: Comparing difference between Cham, Khmer, and Chinese ethnic households in An Giang province, Viet nam. International Journal of Professional Business Review, 7(4), e0527. https://doi.org/10.26668/businessreview/2022.v7i4.e527

Truong, V. D., Hall, C. M., & Garry, T. (2014). Tourism and poverty alleviation: Perceptions and experiences of poor people in Sapa, Vietnam. *Journal of Sustainable Tourism*, 22(7), 1071–1089. https://doi.org/10.1080/09669582.2013.871019.

Trupp, A., Shah, C., & Hitchcock, M. (2023). Globalisation, crafts, and tourism microentrepreneurship in the South Pacific: economic and sociocultural dimensions. *Journal of Heritage Tourism*, 18(6), 733–755. https://doi.org/10.1080/1743873X.2023.2254422.

Wen, S., Cai, X., & Li, J. (2021). Pro-Poor Tourism and Local Practices: An Empirical Study of an Autonomous County in China. *SAGE Open*, 11(2). https://doi.org/10.1177/21582440211022740/ASSET/IMAGES/LARGE/10.1177_21582440211022740-FIG3.JPEG.

Wu, Y., Xu, H., & Eaglen, A. (2011). Tourism-dependent development: the case of Lijiang, Yunnan province, China. *Journal of Policy Research in Tourism, Leisure and Events*, 3(1), 63–86. Http://Dx.Doi.Org/10.1080/19407963.2011.539382.

Yang, J., Wu, Y., Wang, J., Wan, C., & Wu, Q. (2021). A Study on the Efficiency of Tourism Poverty Alleviation in Ethnic Regions Based on the Staged DEA Model. *Frontiers in Psychology*, 12, 642966. https://doi.org/10.3389/FPSYG.2021.642966/BIBTEX.

Yang, L., (Robert) Li, X., & Huang, X. (2023). Ethnic Tourism and Quality of Life: Community Perspectives. In M. Uysal & M. J. Sirgy (Eds.), *Handbook of Tourism and Quality-of-Life Research II* (429–443). Springer. https://doi.org/10.1007/978-3-031-31513-8_29.

12 Navigating the Shadows

A Human Rights-Based Approach to Pro-Poor Mountain Tourism in the Chittagong Hill Tracts

Washik Muhammod Istiaq Ezaz

Introduction

Despite economic progress worldwide, poverty still persists in many mountainous regions. Mountain communities are trapped in cycles of inequalities, have limited opportunities for development and often receive inadequate support from the governmnent. In this context, tourism emerges as a crucial tool for poverty alleviation, providing diverse economic opportunities, especially in impoverished and remote regions with limited alternatives. Pro-poor tourism initiatives that focus on inclusive employment and asset redistribution can play a crucial role in breaking cycles of poverty and inequality. In many mountain regions, poverty is fueled by systemic and structural factors, compounded by uneven economic progress, climate shocks, and prolonged conflicts. Severe poverty, acknowledged as a human rights violation by academics, stresses the urgent need for strategic interventions to address and mitigate these deeply embedded concerns.

Significant challenges, including widespread poverty, vulnerability to natural disasters, and lack of infrastructure, marked Bangladesh's early years. However, political economists argue that Bangladesh defied these challenges through economic growth and pro-poor development policies, pulling millions out of poverty. Closer attention to the GINI index, a measure of income inequality, reveals that aggregate gains may mask uneven progress. In remote mountain areas like the Chittagong Hill Tracts, millions remain trapped in poverty. The region is famous for its picturesque natural beauty but provides limited economic options. The absence of industries and reliance on subsistence agriculture make tourism a promising avenue for generating substantial revenues and alleviating poverty. However, the region's history of protracted conflicts, limited accessibility, and poor administration pose significant obstacles. Combining pro-poor mountain tourism development with implementing human rights can be a feasible solution for overcoming these systemic barriers. A human rights-based approach to pro-poor tourism development will emphasise integrating human rights principles into the planning, implementation, and evaluation of tourism initiatives, focusing on elevating the economic status of the marginalised and poor communities.

DOI: 10.4324/9781003392583-14

This chapter diverges from prior studies concentrating on a monolithic poverty-oriented approach to development cooperation. Instead, it examines the complex interplay between development politics and human rights within the conflict-ridden historical context of the region. Specifically, it investigates how the region's systematic deprivation, rooted in existing legal frameworks, perpetuates historical marginalisation. By proposing a human rights-based approach to pro-poor mountain tourism development as a potential pathway, this chapter explores sustainable development possibilities that prioritise tribal communities and their unique socio-economic needs.

Theoretical Background

The Human Rights-Based Approach

In tourism human-rights-based approach is predominantly unutilised and only some countries have already incorporated this approach into their tourism policies and practices. This conceptual framework for human development is fused with international human rights standards, aiming to champion and safeguard those rights by linking their normative basis to feasible implementation (Broberg & Sano, 2017). The human-rights-based approach involves core principles, including accountability, the rule of law, participation and inclusion, and equality and non-discrimination. This translates into a methodological framework for implementing human rights within development contexts, specifically addressing inequalities that hinder pro-poor development (Vandenhole & Gready, 2014). This framework prevents socio-economic inequalities through empowering marginalised communities, ensuring equitable economic opportunities, and addressing biased land-use policies. Notably, a human rights-based approach helps to recalibrate the core mission of development from mere charity to an obligation to respect, protect, and fulfil rights. While existing development models encourage participation, inclusion and economic incentivisation, they often overlook underlying human rights concerns like intersectional discrimination and land ownership that are crucial for marginalised communities (Ashley et al., 2001). In essence, the human rights-based approach bridges the gap in traditional development models by addressing systemic inequalities at their core, thereby empowering communities and advancing sustainable development.

Human Rights-based Approach for Inclusive Pro-Poor Tourism Development in Chittagong Hill Tracts

The core international human rights treaties do not explicitly use poverty alleviation in their legal instruments. However, it includes provisions relevant to the broader concept of improving living standards and ensuring a decent quality of life for all individuals. Through monitoring and reporting mechanisms, human rights instruments highlight the detrimental impact of poverty on various fundamental rights and emphasise the responsibility of nation-states to work towards its alleviation. It is often argued that extreme poverty deprives individuals of their fundamental human rights, and global poverty is not just a result of natural disasters or individual failures but a consequence of unjust global institutional arrangements (Pogge, 2005). Individuals and institutions in affluent nations and the global system bear a moral responsibility for addressing and alleviating the extreme poverty suffered by a significant part of the world's population. Despite initial scepticism after independence, Bangladesh has achieved remarkable progress in alleviating poverty. Through macroeconomic stability and trade liberalisation, capacity for managing natural disaster vulnerability, and targeted pro-poor development policies, millions have escaped the grips of poverty (Besley & Cord, 2007). External factors like international aid and development models played a significant role, too (Hossain, 2017). However, the core of this success story stems from homegrown initiatives tailored to Bangladesh's specific context (Roy et al., 2016). Despite consistent growth in per capita income that surpasses many neighbouring countries, a closer introspection reveals a rather alarming trend. The Gini Coefficient, which measures income inequality, has risen sharply to 0.499 (Bangladesh Bureau of Statistics, 2023). This underscores the challenge of an uneven distribution of economic gains despite overall growth, which is further exacerbated in the isolated Chittagong Hill Tracts due to a lack of infrastructure, resources and the interplay of ancillary factors like ethnicity, religion, and gender. This complicated interaction manifests in alarmingly high poverty rates and limited access to necessities for marginalised communities (Alam & Iqbal, 2017). The ongoing situation has proved that a poverty-oriented approach falls short of addressing the unique challenges experienced in this particular region.

Tribal communities in the Chittagong Hill Tracts rely heavily on traditional practices like Jhum cultivation, hunting, and crafts, but these offer limited income and are lashed with increasing challenges. Production, marketing and supply chain systems are often inaccessible, leaving the hill population in economic despair. Tourism can be a potential lifeline for them, offering diverse jobs, entrepreneurial opportunities, and improved infrastructure. Many scholars argue that a capability approach based on human rights principles can be valuable for the alleviation of poverty. The capability approach, emphasising the expansion of freedoms for well-being, resonates with the needs of tribal communities in the Chittagong Hill Tracts, who cannot often participate in mainstream economic activities of the society (Atkinson, 2019). However, due to intersecting inequalities based on ethnicity, religion, and gender, a human rights-based approach that addresses underlying structural inequalities, respects diverse cultures and land rights, empowers local communities, and prioritises culturally sensitive initiatives could be a viable solution for this region. This aligns with the growing interest in mountain tourism, which can be a powerful tool for poverty alleviation when coupled with human rights principles. However, concerns remain about objectification and undermining local autonomy (Freire-Medeiros, 2014). Through recognising the Chittagong Hill Tracts' unique dynamics, complex administrative systems, and history of protracted conflict, this approach addresses systematic marginalisation and prioritises respecting the host community's socio-economic rights.

Methodology

This research explores the feasibility and significance of a human rights-based approach to pro-poor mountain tourism development in Bangladesh's Chittagong Hill Tracts (CHT). It aims to understand how a human rights-based approach can address systemic barriers and promote sustainable development that prioritises the rights and needs of marginalised tribal communities. This research integrates socio-legal human rights perspectives to thoroughly examine the feasibility and significance of a human rights-based approach in developing pro-poor mountain tourism in Chittagong Hill Tracts. The research adopts a qualitative approach rooted in a theoretical and analytical knowledge framework in law, society and politics. The methodology comprises two principal components. First, a doctrinal step involves identifying and analysing legal sources, employing methods synonymous with document and content analysis (Hutchinson & Duncan, 2012). Second, a sociological step interprets the law within the context of the research topic, drawing on historical and ethnographic studies of the Chittagong Hill Tracts. This dual-step approach ensures a comprehensive understanding of the legal landscape, power dynamics, and traditional governance structures within tribal communities, allowing for a nuanced exploration of the research questions (Creutzfeldt et al., 2019).

Trapped in the Shadow of Conflict: Unveiling Chittagong Hill Tracts

The Chittagong Hill Tracts, in southeast Bangladesh bordering Myanmar (Burma) and India, comprise three hill districts. This ethnically and culturally diverse region is home to eleven tribal groups, including the *Chakma, Marma, Tripura*, and *Tanchangya*, each with distinct cultures, languages, and religions (Mohsin, 2003). The area is characterised by steep, rugged mountainous terrain, dense forests, and tribal groups living in scattered cluster habitats that are difficult to access. The politico-legal history of the region is characterised by a convoluted legacy of colonisation and conflict (Mohsin, 1997). With the Chittagong Hill Tracts Regulation of 1900, the British colonial rulers facilitated special protection for the tribal population and their land rights. Subsequently, the leadership of the CHT rejected the political process leading to the partition of the Indian Subcontinent in 1947 and resisted the armed struggle that gave birth to the Independent Bangladesh (erstwhile East Pakistan) in 1971. Their failure to comprehend the evolving political dynamics led to the meltdown of law-and-order situations in this region, exacerbated through protracted insurgency and militarisation as countermeasures (Badiuzzaman & Murshed, 2015). This violent conflict in the CHT consequently affected its tourism potential. The 1997 CHT Peace Accord signed between the Government of Bangladesh and the *Parbatya Chattagram Jana Sanghati Samiti* (United People's Party of the Chittagong Hill Tracts) was a significant step towards peace and development. While the Accord brought peace for the time being, a few rebel groups did not comply with the peace process, and tensions occasionally arose. The Bangladeshi government maintains a considerable security presence in the region, which is a deterrent to large-scale armed activity.

The long-standing political instability of the CHT area, exacerbated by violent insurgency, has resulted in the region being excluded from mainstream development processes. Agriculture is the primary economic activity, given the region's topography and climatic conditions. The region's remoteness and limited accessibility have constrained its ability to attract investment and sustainable economic growth. Consequently, the region faces significant socio-economic challenges, with most of its population living below the poverty line, surpassing the national average poverty rate (Bangladesh Bureau of Statistics & World Food Programme, 2020). This distressing condition has a detrimental effect on socio-economic progress, aggravating the region's underdevelopment.

Colonial Legacy to Modern Challenges: Socio-Economic and Legal Vulnerabilities

Displacement and jeopardized land rights

Chittagong Hill Tracts has experienced displacement and complicated land rights issues, which can be traced back to the British colonial era. These challenges stem from a complicated history of conflicts, settlement policies, and struggles for autonomy. Consequently, land rights have become a critical concern for the tribal communities in the region (Yasmin, 2014). The British colonial government established a system of indirect rule in Chittagong Hill Tracts in the early 19th century. This indirect governance helped to preserve the traditional way of life for the indigenous peoples, but it also created a system of dependency that made them vulnerable to exploitation (Shahabuddin, 2018). The colonial government implemented a policy of isolation, which aimed to prevent the hill people from claiming any form of autonomy and exclude them from representation in Bengal politics. This policy had detrimental consequences, leading to the people of CHT being unrepresented at both provincial and national levels (Chowdhury, 2016). As a result, they could not voice their opinions on the partition question since different vested groups lobbied unsuccessfully to exclude Chittagong Hill Tracts from

Muslim-majority East Pakistan. All this happened even though economic ties, historical relations and geopolitical cohesiveness completely bound the region with East Pakistan. Right after the partition of the Indian Sub-continent, Chittagong Hill Tracts emerged as a region of significant geostrategic interest, mainly due to its location on the eastern flank. The Pakistani government adopted a comprehensive policy of militarisation in the region and revoked the exclusionary political status previously granted by the British colonial rulers. Additionally, government policies encouraging Bengali settlement and infrastructure development initiatives attracted Bengali settlers to the region. The construction of the *Kaptai* dam in the 1960s triggered large-scale displacement and migration of the native population and the submersion of tribal lands.

The settlement process escalated after the Bangladesh Liberation War in 1971. The conflict and subsequent political changes contributed to demographic shifts, with an increased Bengali population in the region. The Bengali settlement has been one of the triggering factors of tension and conflict among the tribal communities. This has led to a protracted conflict marked by the demand for autonomy and the recognition of the rights of the tribal communities., the insurgency and the responding military operations led to widespread violations of human rights in the region (Acemoglu & Robinson, 2012). Eventually, through a protracted negotiation process, the Bangladesh government managed to secure a peace process in 1997, which ended the conflict and the discriminatory settlement policy, facilitating the repatriation of the displaced population from India (Haidar, 2006). The peace accord facilitated the recognition of tribal customs and land rights and the establishment of a regional council with limited self-governance authority. However, the full implementation of the Accord has been slow, and numerous challenges persist, including land disputes, unequal political representation, and unplanned socio-economic development in the region (Adnan & Dastidar, 2011). A specialised land commission was ultimately established to safeguard tribal land rights and settle ongoing disputes. However, a significant challenge stems from the region's communal history of land ownership, as the hill people primarily lack ownership deeds.

Legal and institutional frameworks

The Constitution of Bangladesh has incorporated provisions to ensure preferential treatment for disadvantaged segments of the population, which includes tribal groups and ethnic minorities. In 2011, the constitution was amended by inserting Article 23A, which addresses explicitly protecting and developing the unique local culture and traditions of tribes, minor races, ethnic sects, and communities (Roy, 2011). This recognition of indigenous peoples as a distinct group within the Constitution of Bangladesh fosters trust and cooperation between mainstream society and tribal communities, particularly in the aftermath of the peace accord. The term 'minor races' in this carefully crafted acknowledgement has been met with concern by many human rights organisations and stakeholders. The Government of Bangladesh maintains its ardent opposition to recognising tribal populations as indigenous peoples in its constitution and, instead, prefers to use the terms "tribe" and "tribal" rather than "indigenous".

The International Labour Organization has formulated two global instruments related to indigenous people. While Bangladesh ratified the more general Indigenous and Tribal Populations Convention (No. 107) in 1972, the region's development calls explicitly for the comprehensive protections outlined in the unratified Indigenous and Tribal Peoples Convention, 1989 (No. 169) (International Labour Organization, 2009). This Convention's emphasis on securing land rights, meaningful participation in decision-making, and safeguarding cultural heritage aligns with alleviating historical marginalisation and promoting sustainable development in the region.

The Peace Accord was signed to ensure the active participation of the tribal population within the governance of Chittagong Hill Tracts. This Accord marked a significant milestone in recognising the internal autonomy of the region and seeking to restore its special status. As part of this effort, a three-tier administrative system was introduced for CHT, accompanied by the recognition of the need to enact new laws, amend the existing ones, and establish regulations and practices that would ensure compliance with the provisions of the Peace Accord. In response to this requirement, the government introduced several legislative initiatives to address the concerns and demands outlined in the Peace Accord. These initiatives included the enactment of the Chittagong Hill Tracts Regional Council Act of 1998, the Chittagong Hill Tracts Land Disputes Resolution Commission Act of 2001, and the amendment of the Chittagong Hill Tracts Regulation Act in 2003. Notably, amendments were made to the Hill District Council Act 1998 to make them comply with the provisions of the Peace Accord. These amendments granted the council increased authority and autonomy by expanding the range of subjects under their jurisdiction.

A tourism policy's indispensable role in guaranteeing a destination's success is undeniable since it offers a comprehensive guide to safeguard the interests of destination stakeholders while minimising negative impacts. Unfortunately, the National Tourism Policy 2010 falls short of recognising the distinctiveness and tourism potential of the Chittagong Hill Tracts, with minimal mention of Chittagong Hill Tracts in the discussion of "nature-based" and "cultural" aspects of tourism (Rahman et al., 2020).

Barriers to the implementation of legal and institutional frameworks

The constitutional recognition provided to the tribal population under Article 23A is embedded within the fundamental principles of policy, which are essentially non-justiciable. Moreover, the implementation of the Peace Accord lacks constitutional recognition (Sharmin, 2014). In order to preserve the integrity of the Accord and other relevant statutes, it is suggested that their inclusion in the Constitution's schedule would afford them constitutional protection. Furthermore, stakeholders perceive a lack of necessary political will regarding implementing the Accord. Within the framework of the Accord, the Ministry of Chittagong Hill Tracts Affairs, Chittagong Hill Tracts Regional Council, and Land Commission were established to address different governance and resource allocation issues. Despite achieving intermittent advancements, tribal communities have expressed substantial discontent with the Peace Accord's overall outcomes. Findings from a study conducted by the National Human Rights Commission indicate that only 48 of the 72 objectives have been implemented to date (National Human Rights Commission, Bangladesh, 2014). The Accord has a distinctive status, as it neither derived from Parliament nor qualifies as an international treaty under the Vienna Convention on the Law of Treaties, 1969. The primary factors contributing to the limited progress can be attributed to the hierarchical nature of the Accord, which lacks informed consent from the citizens and consequently disregards traditional power structures within Chittagong Hill Tracts. These factors, in turn, create significant barriers to further development in that region.

With the heightened security concerns, the unresolved issue of land rights and Bengali settlement continues to generate tension and remain relevant within the region. Notably, a significant proportion of the tourists were initially expected to comprise foreign nationals. However, all foreign nationals, including those of Bangladeshi origin and having dual citizenship, are currently required to have prior permission before visiting Chittagong Hill Tracts. This requirement has allowed bureaucratic red tape, thus impeding the gradual development of tourism in the region.

From sanctuary to souvenir: Chittagong Hill tracts in tourism's crossroads

Tourism has the potential to alleviate poverty and contribute to the economic well-being of local communities by creating employment opportunities and generating income, particularly in areas out of reach for conventional development initiatives. Since the mountains are characterised as inaccessible, fragile, diversified and marginal areas, the region possesses a distinctive and delicate ecosystem, requiring tourism practices that consider both local communities and the environment (Jones et al., 2021). Unfortunately, tourism in the Chittagong Hill Tracts region is growing without proper attention to environmental and cultural concerns. The absence of conservation initiatives, regulations, and adequate supervision in the tourism sector has resulted in adverse effects on the environment, wildlife, and the traditional practices of tribal communities (Apollo, 2021).

Tourists visit the region to experience the exotic lifestyle of tribal communities and indulge in the aesthetic beauty of its unparalleled landscape, often without showing much regard for the local communities and environment.

One top-rated destination is Sajek, where the creation of tourist spots has displaced hill residents from specific areas, shifting their value from cultural significance to primarily market appeal. This commercialisation of nature signifies a fundamental change in the meaning of the environment, transforming it from a source of direct sustenance to a commodity with exchange value (Sajib, 2021).

Furthermore, frequent landslides during the monsoon season pose a significant threat in the region. Unplanned development initiatives, such as sand and stone extraction and road construction aimed at facilitating tourism, directly and indirectly, contribute to the triggering of landslides (Ahmed et al., 2022). It is concerning that there is a lack of detailed plans for the tourism industry and specific construction policies tailored to the rugged landscape of the region. Tourism development policies implemented in the region often overlook the needs and interests of the local communities. Consequently, these communities are marginalised, deprived of the benefits of tourism, and left with their resources being exploited by external actors without fair profit-sharing mechanisms.

As a result, tourism in the region is seen as a form of soft domination over ethnic minorities despite its promoted role in fostering local community development. The unplanned tourism development in the region has damaged the sense of belonging and the reciprocity between tribal communities and their natural environment. The commercialisation of culture and environment has not only altered their perception of cultural and natural resources but also disrupted their everyday way of life. However, the logical solution to this dilemma would be creating a symbiotic relationship wherein tourism activities within the area are conducted in a manner that ensures environmental and social sustainability.

Unravelling Legal and Policy Insights: Themes and Patterns in Chittagong Hill Tracts Context

The first theme is displacement and land rights violations. Historical displacement, beginning from the British colonial era and exacerbated by the construction of the Kaptai dam and settlement policies, has led to the submersion of tribal lands and forced migration. Bengali settlement after the Bangladesh Liberation War has been a source of tension and conflict, contributing to a protracted demand for autonomy and recognition of tribal rights. Land disputes persist, mainly due to the communal nature of land ownership, creating barriers to recognising tribal women's land rights. Another theme is insurgency and militarisation impact. The conflict and subsequent political changes, including the insurgency and military operations, resulted in widespread human rights violations in the Chittagong Hill Tracts. Though a peace accord emerged, challenges remain, including slow implementation, political representation issues, and socio-economic development concerns. The next theme concerns political and administrative challenges. The hierarchical nature of the Peace Accord, lack of informed consent, and exclusion of traditional power structures have hindered its implementation and contributed to discontent among tribal communities. Limited progress is attributed to a lack of political will, as a significant number of objectives outlined in the Accord have not been implemented, according to a study by the National Human Rights Commission. Security measures and tourism barriers also play an essential role in the context of Chittagong Hill Tracts. More rigid security rules in the area affect tourism because of safety worries and new rebel groups. Also, rules for foreign visitors are making it harder for tourism to grow.

Furthermore, the region is characterised by inadequate legal and institutional frameworks. The constitutional recognition enshrined in Article 23A lacks a mechanism for judicial enforcement, limiting its practical application. Similarly, the Peace Accord lacks the legal weight of constitutional protection, potentially hindering its full implementation. Stakeholders perceive a deficiency in the political will regarding the implementation of the Accord, which creates significant obstacles to future advancements and developments. Finally, in mountain regions, it is essential to consider the impact of tourism development and its relation to cultural disruption. The absence of careful planning in tourism development can lead to adverse environmental effects, with landslides becoming more prevalent. Moreover, transforming nature and culture into commodities disrupts the intrinsic connection that tribal communities have with their surroundings. This disruption can have far-reaching consequences as it severs the traditional bond between these communities and their natural environment, potentially leading to a loss of heritage and identity. Therefore, it is crucial to approach tourism development in these regions with a focus on sustainability and preserving the cultural integrity of the local communities.

Beyond Mountains and Tribes: Integrating a Human Rights-Based Approach in Pro-Poor Mountain Tourism

Tourism, when appropriately applied, can serve as an effective policy tool for poverty reduction (Lagos & Wang, 2023). However, evidence indicates that the tribal communities in the Chittagong Hill Tracts grapple with intricate challenges that a conventional, poverty-oriented approach to development has proved inadequate to meet (Datta, 2019). The region's social accessibility barriers, stemming from its status as a border region, conflicts, administrative travel restrictions, and various forms of unrest, collectively hinder accessibility and openness to tourism (Apollo & Wengel, 2022). Acknowledging these obstacles, adopting a human rights-centered approach is viable and crucial for fostering comprehensive and enduring development. The region's history of conflict, displacement, and flagrant human rights violations demands a comprehensive framework that rectifies past injustices. By leveraging existing legal and institutional frameworks, a human rights-based approach can effectively implement the Peace Accord, safeguarding indigenous rights and culture.

Integrating human rights into tourism planning fosters sustainability, respects cultural practices, and ensures the local community's benefit. Ultimately, a human rights-based approach offers a holistic and transformative approach to pro-poor mountain tourism development in the region, emphasising the crucial role of human rights in building a resilient and thriving community.

Conclusion

The Chittagong Hill Tracts present a complex landscape where historical injustices, displacement, and socio-economic challenges intersect. This research underscores the potential of a human rights-based approach to pro-poor mountain tourism, recognising economic needs and social justice issues with a nuanced understanding of individual and group experiences. While promoting participation and offering economic benefits, existing development models often neglect the crucial human rights dimension of empowering marginalised communities. The exclusive emphasis on poverty alleviation, while neglecting fundamental human rights issues such as intersectionality and land ownership, has clearly demonstrated its inadequacy in previous years. As a result, a crucial shift towards a human rights-based approach becomes imperative as a critical perspective for redefining pro-poor development in the region. Hence, this research advocates for a comprehensive re-evaluation of tourism policies, legal frameworks, and development practices through human rights vernacular, fostering a symbiotic relationship between economic growth, community empowerment and cultural preservation.

Management Implications

In fostering inclusive and sustainable tourism development, stakeholders and policymakers need to collaborate and prioritise creating inclusive policies that consider the intersecting needs of diverse communities. Achieving this goal demands a thorough understanding of historical injustices and a dedicated effort to rectify them, guaranteeing that the unique needs of various demographic groups are recognised and effectively addressed. Additionally, investing in tourism education and awareness programs is crucial, targeting both host communities and tourists. Responsible and sustainable tourism practices can be encouraged by educating visitors about the cultural and environmental significance of the regions they explore.

References

Acemoglu, D., & Robinson, J. A. (2012). *Why nations fail: The Origins of Power, Prosperity and Poverty.* Profile Books.

Adnan, S., & Dastidar, R. (2011). *Alienation of the Lands of Indigenous Peoples in the Chittagong Hill Tracts of Bangladesh.* Chittagong Hill Tracts Commission and International Work Group for Indigenous Affairs.

Ahmed, B., Alam, S. M. R. A., Ahmed, I., & Sammonds, P. (2022). The Anthropogenic Aggravation of Landslide Disasters in Bangladesh: Key Informants' Perspectives. In I. Alcántara-Ayala, Ž. Arbanas, D. Huntley, K. Konagai, M. Mikoš, K. Sassa, S. Sassa, H. Tang, & B. Tiwari (Eds.), *Progress in Landslide Research and Technology* (385–401). Springer. https://doi.org/10.1007/978-3-031-18471-0.

Alam, S., & Iqbal, K. (2017). Dynamics of regional poverty and real wages: Policy implications for development interventions. In J. Devine, G. D. Wood, Z. Ali, & S. Alam (Eds.), *Extreme Poverty, Growth, and Inequality in Bangladesh* (217–259). Practical Action Publishing Ltd. https://doi.org/10.3362/9781780449463.006.

Apollo, M. (2021). *Environmental Impacts of Mountaineering: A Conceptual framework.* Springer. https://doi.org/10.1007/978-3-030-72667-6.

Apollo, M., & Wengel, Y. (2022). *Mountaineering tourism: A critical perspective.* Routledge. https://doi.org/10.4324/9781003095323.

Ashley, C., Roe, D., & Goodwin, H. (2001). *Pro-poor tourism strategies: Making tourism work for the poor: A review of experience.* Overseas Development Institute.

Atkinson, A. B. (2019). *Measuring Poverty around the World.* Princeton University Press https://doi.org/10.2307/j.ctvc77fd6.

Badiuzzaman, M., & Murshed, S. M. (2015). Conflict and livelihood decisions in the Chittagong Hill Tracts of Bangladesh. In A. Heshmati, E. Maasoumi, & G. Wan (Eds.), *Poverty Reduction Policies and Practices in Developing Asia* (145–164). Asian Development Bank and Springer. https://doi.org/10.1007/978-981-287-420-7.

Bangladesh Bureau of Statistics & World Food Programme (2020). *Poverty Maps of Bangladesh 2016.*

Bangladesh Bureau of Statistics (BBS) (2023). *The Household Income and Expenditure Survey (HIES).*

Besley, T., & Cord, L. J. (2007). *Delivering on the Promise of Pro-Poor Growth: Insights and Lessons from Country Experiences.* World Bank and Palgrave Macmillan.

Broberg, M., & Sano, H. (2017). Strengths and weaknesses in a human rights-based approach to international development – an analysis of a rights-based approach to development assistance based on practical experiences. *The International Journal of Human Rights*, 22(5), 664–680. https://doi.org/10.1080/13642987.2017.1408591.

Chowdhury, T. M. (2016). *Indigenous Identity in South Asia: Making Claims in the Colonial Chittagong Hill Tracts.* Routledge. https://doi.org/10.4324/9781315561967.

Creutzfeldt, N., Mason, M., & McConnachie, K. (2019). *Routledge Handbook of Socio-Legal Theory and Methods.* Routledge. https://doi.org/10.4324/9780429952814.

Datta, R. (2019). Implementation of Indigenous environmental heritage rights: An experience with Laitu Khyeng Indigenous community, Chittagong Hill Tracts, Bangladesh. *AlterNative: An International Journal of Indigenous Peoples*, 15(4), 309–320. https://doi.org/10.1177/1177180119885199.

Freire-Medeiros, B. (2014). *Touring poverty.* Routledge. https://doi.org/10.4324/9780203840719.

Haidar, N. (2006). Indigenous Peoples of Bangladesh and the Chittagong Hill Tracts Peace Accord: A case of Distrust and Disintegration. *The Dhaka University Studies, Part-F,* XVII(1), 25–48.

Hossain, N. (2017). *The Aid Lab.* Oxford University Press. https://doi.org/10.1093/acprof:oso/9780198785507.001.0001.

Hutchinson, T., & Duncan, N. (2012). Defining and describing what we do: Doctrinal legal research. *Deakin Law Review*, 17(1), 83. https://doi.org/10.21153/dlr2012vol17no1art70.

International Labour Organization (2009). *The ILO Convention on Indigenous and Tribal Populations, 1975 (No. 107) and the Laws of Bangladesh: A Comparative Review.*

Jones, T. E., Apollo, M., & Bui, H. T. (2021). Mountainous Protected Areas & Nature-Based Tourism in Asia. In T. E. Jones, H. T. Bui, & M. Apollo (Eds.), *Nature-Based Tourism in Asia's Mountainous Protected Areas: A Trans-regional Review of Peaks and Parks* (pp. 3–25). Springer. https://doi.org/10.1007/978-3-030-76833-1.

Lagos, K., & Wang, Y. (2023). International tourism and poverty alleviation: Cross-country evidence using panel quantile fixed effects approach. *Journal of Travel Research*, 62(6), 1347–1371. https://doi.org/10.1177/00472875221119978.

Mohsin, A. (1997). *The Politics of Nationalism*. The University Press Limited.

Mohsin, A. (2003). *The Chittagong Hill Tracts, Bangladesh: On the difficult road to peace*. Lynne Rienner Publishers.

National Human Rights Commission, Bangladesh. (2014). *Implementation of the Chittagong Hill Tracts Peace Accord: Challenges and Human Rights Issues.*

Pogge, T. (2005). World Poverty and human Rights. *Ethics & International Affairs*, 19 (1), 1–7. https://doi.org/10.1111/j.1747-7093.2005.tb00484.x.

Rahman, M. S.-U., Faisal, A., Hassan, A., & Hoque, M. A. (2020). Tourism Policy in Bangladesh: A Critical Perspective. In M. S.-U. Rahman & A. Hassan (Eds.), *Tourism Policy and Planning in Bangladesh* (pp. 641–673). Springer. https://doi.org/10.1007/978-981-15-7014-8_18.

Roy, A., Negrón-Gonzales, G., Opoku-Agyemang, K., & Talwalker, C. (2016). *Encountering Poverty: Thinking and Acting in an Unequal World*. University of California Press.

Roy, R. D. (2011). State Policy, Constitution and Equal Rights for Indigenous Peoples' of Bangladesh. In P. Gain (Ed.), *Survival on the Fringe: Adivasis of Bangladesh* (497–502). Society for Environment and Human Development (SEHD).

Sajib, S. M. S. A. (2021). Nicknaming tourism as development: commercialisation of culture and nature in CHT, Bangladesh. *Journal of Tourism and Cultural Change*, 20(1–2),279–290. https://doi.org/10.1080/14766825.2021.1966022.

Shahabuddin, M. (2018). The Myth of Colonial 'Protection' of Indigenous Peoples: The Case of the Chittagong Hill Tracts under British Rule. *International Journal on Minority and Group Rights*, 25(2), 210–235, https://doi.org/10.1163/15718115-02502008.

Sharmin, T. (2014). Constitutionality of the Chittagong Hill Tracts Peace Accord 1997: Is the Peace Process Valid? *Law Asia Journal*, 1–26. http://doi.org/10.2139/ssrn.3171276.

Vandenhole, W., & Gready, P. (2014). Failures and Successes of Human Rights-Based Approaches to Development: Towards a Change Perspective. *Nordic Journal of Human Rights*, 32(4), 291–311. https://doi.org/10.1080/18918131.2015.957458.

Yasmin, L. (2014). The Tyranny of the Majority in Bangladesh: The Case of the Chittagong Hill Tracts. *Nationalism and Ethnic Politics*, 20(1), 110–120. https://doi.org/10.1080/13537113.2014.879769.

Part III

Lessons from Advanced Economies

13 Tourism's Role in Addressing Poverty through Asset Equity in the U.S. Mountain West

Bryan Breidenbach and David W. Knight

Introduction: Sustainable Tourism Development and Affordable Housing

For decades the tourism industry has become accustomed to annual growth rates measured by tourist departures, arrivals, and other various economic indicators. Its impact is evident as the tourism industry directly involves or supports one in ten jobs globally (WTTC, 2017). In the United States alone, tourism directly employed over six million individuals in 2019. However, an estimated half of the tourism jobs were lost in the United States in 2020 in response to public health measures related to the COVID-19 pandemic (IBIS World, 2022). Similarly, global economic data indicates that nearly five trillion dollars were lost in tourism revenues in 2020 (WTTC, n.d.). As COVID-19-related health measures relaxed, tourism job creation and revenues rebounded but have not yet reached pre-pandemic levels (UNWTO, 2022). People employed in the tourism industry are uniquely vulnerable to negative livelihood outcomes in response to shocks such as the COVID-19 pandemic, in part because tourism-related jobs often do not require advanced skill sets (McCool & Bosak, 2015; Scheyvens & Hughes, 2019), frequently offer low employment wages (Darling, 2005; Schively, 2008) and are labour intensive (Winter & Kim, 2021). These factors create conditions that have resulted in questioning whether tourism is a sustainable cure to poverty, as claimed by the Sustainable Development Goals of the United Nations World Tourism Organization (UNWTO) (Scheyvens & Hughes, 2019).

Poverty alleviation is a primary goal of the global tourism industry and associated tourism organisations, such as the UNWTO, through the creation of opportunities via economic development (Álvarez-Herranz, & Macedo-Ruíz, 2021; Knight et al., 2017). Communities dependent on tourism have clear incentives to ensure resources are managed responsibly, thus ensuring predictable future tourism growth and local economic opportunities (Ooi et al., 2016). To aid long-term destination planning, the United Nations World Commission on Environment and Development (UNWCED) popularised the concept of sustainable development, defining sustainable development as "development that meets the needs of the present without compromising the ability of future generations to meet their own needs" (UNWCED, 1987, p. 41).

DOI: 10.4324/9781003392583-16

Subsequent UNWCED reports elaborated three essential elements, or pillars, of sustainable development, which include environment, economy, and social equity (UNWCED, 1992). The pillars of sustainable development are frequently described as conflicting with one another due to perceived tradeoffs regarding the goals of sustainable development (Campbell, 1996; Frederick, 2022; Schively, 2008). Sustainable development is a complex social-ecological system requiring community agency and resilience through adaptive capacity to achieve sustainable development goals (Ruiz-Ballesteros, 2011). Resilience is "the capacity of a system to absorb disturbance and reorganise while undergoing change to retain essentially the same function…" (Walker et al., 2004, p. 7).

Campbell (1996) illustrates the complex decisions that sustainable development planners face as a "planner's triangle", where social equity conflicts with both economic (property conflict) and environmental (development conflict) considerations, while economic and environmental components experience competing conflict over resources. Conflicts associated with achieving goals sought by communities implementing sustainable development are frequently rooted in psychological and social-political-economic issues (Almeida et al., 2017). Managing sustainable development conflicts requires an additional institutional dimension of sustainable development, as Cottrell et al. (2013) described. Institutions within the prism of sustainability can be considered governance or decision-making bodies with the authority and responsibility to address conflicts associated with perceived tradeoffs related to managing the environment, economy, and equity pillars of sustainable development (Cottrell et al., 2013). These sustainable development concepts and frameworks can be used to inform sustainable tourism, which is defined by the UNWTO (n.d.) as "tourism that takes full account of its current and future economic, social, and environmental impacts, addressing the needs of visitors, the industry, the environment, and host communities" (Yiu, & Cheung, 2021, p. 2). This review is focused on applying these concepts to better understand if sustainable tourism in the Mountain West of the United States can achieve sustainable poverty alleviation.

Any effort to address equity and poverty alleviation as crucial pillars of sustainable development requires clarity around how poverty is defined. This study follows Narayan et al. (2004), who interviewed 40,000 persons experiencing poverty to better understand their perspectives. Their research identified multiple dimensions to poverty beyond income, defining it as a "lack of what is necessary for material well-being – including housing" while linking "their lack of physical, human, social, and environmental assets to their vulnerability and exposure to risk" (Narayan et al. 2004, p. 31). Measuring poverty by asset capability rather than income clarifies how sustainable tourism may help destinations pursue and potentially achieve social equity. This definition implicitly points to affordable housing as an essential contributor to poverty alleviation. The U.S. Department of Housing and Urban Development (HUD) (2017) defines affordable housing as a cost of thirty per cent or less of a household's monthly income.

During the twentieth century, many Mountain West communities in the United States have transitioned from economies based on resource extraction to tourism-dependent economies (Amason, 2020; Baron et al., 2000). Mountain West tourism communities are often characterised by geographic isolation and conservation-focused land use policies that restrict development (Brooks, 2022; Schively, 2008). These communities frequently face housing demand pressures exceeding existing housing supply, creating imbalances in housing market (Amason, 2020). The impacts of the COVID-19 crisis have resulted in increased attention from industry and academia on how resilience thinking can improve sustainable tourism development (Yi & Cheung, 2021).

In view of these concerns and trends, the purpose of this chapter is to explore the affordable housing crisis that manifested in many tourism destinations in the Mountain West prior to and during the COVID-19 pandemic (Bell Policy Center, 2022). It explores key themes in this context, including the role of tourism as a mechanism for poverty alleviation, the sustainable tourism development framework to address growing property inequity, how resilience thinking can address shocks related to affordable housing, and the psychology of motivations influencing stakeholders to address affordable housing. The themes of tourism, poverty alleviation, sustainable tourism development, resilience, and motivations are coupled in the complex social-ecological-systems of destinations reliant on tourism-related economic growth, as decision-making related to one influences and is influenced by the others. Holling (2001) provides additional clarity to sustainable development in the context of social-ecological-systems by defining sustainability as "the capacity to create, test, and maintain adaptive capability" and development as "the process of creating, testing, and maintaining opportunity" (p. 390). By exploring these relationships, a better understanding can be gained of destinations" capacities to create positive outcomes for those experiencing poverty and who are the most vulnerable to system shocks in the Mountain West.

Background: Social Equity in Sustainable Tourism Development

The plight of many Mountain West communities in the mid-twentieth century was not unique. As previous well-paying resource extraction jobs disappeared, so too did populations through out-migration, resulting in depressed economies throughout the Appalachia region (Amason et al., 2020). Many Mountain West communities sought to increase economic resilience by developing tourism industries to create in-migration, local economic opportunities, and enhanced infrastructure (Ibanescu et al., 2020). Beginning in the 1980s to the present, in-migration began to change Mountain West communities, as an influx of affluent developers and individuals began threatening traditional western livelihoods (Baron et al., 2000). Livelihoods can be understood as "capabilities, assets, and activities required for a means of living" (Scoones, 1998, p. 5). Motivations for migration are inherently complex from a macro-level context (Weaver & Lawton, 2017). The in-migration being experienced in the Mountain West and other rural communities has been associated with out-migration from urban areas, second homes or investment property acquisition, and access to nature and recreation opportunities (Darling, 2005). Today, Mountain West communities are largely reliant on tourism for economic opportunities, which has created conflict in relation to the environmental, economic, and equity pillars of sustainable development, and debate continues on whether these pillars were ever co-equals (Becker et al., 2015). The early stages of euphoric tourism development will likely proceed to annoyance and antagonism on behalf of host communities and visitors (Apollo & Andreychouk, 2022). Ensuing conflict and perceived tradeoffs between the pillars of sustainable development are most capable of being resolved at an institutional level (Cottrell et al., 2007). This institutional lens is applied to the following thematic sections regarding decision-making to address the affordable housing crisis experienced in tourism-dependent economies of the Mountain West.

Tourism as a Tool to Address Poverty

Tourism is frequently billed as an industry capable of alleviating poverty. This claim is often measured in terms of monetary income based on employment opportunities (Winter & Kim, 2021). Such narrow thinking about poverty and associated relief through employment opportunities fails to recognise the multi-dimensional aspects of poverty, which should be further defined beyond income metrics (Becker et al., 2015). Context is also important when considering the multi-dimensional nature of poverty, as socio-demographic characteristics such as gender, age, education, and geographic location influence how poverty is both perceived and defined by people experiencing poverty (Narayan et al., 2004). Knight et al. (2017) investigated tourism as a means of poverty alleviation in Peru by applying the pro-poor tourism framework to the local context. Pro-poor tourism consists of "strategies that increase the benefits to poor people from tourism development" (Ashley & Roe, 2020, p. 61). Knight (2018) focused on pro-poor tourism to promote local development by including and promoting local community perceptions to influence the development process via institutional channels, or the degree to which locals can participate and influence sustainable tourism development in host communities. This is an important aspect of the prism of sustainability, which asserts that the pillars of sustainable development (environment, economy, equity) maintain credible sustainability only when locals feel involved in the planning, managing, and policy of destination resources for equitable growth opportunities (Cottrell et al., 2007).

While there are limitations to comparing local perceptions of poverty in mountain communities of Peru and the U.S., there are similarities between the two. The qualitative data from Peru, for example, suggest that locals view tourism as a means of poverty alleviation through employment but question whether tourism assets, such as a newly developed airport, would be beneficial to communities" well-being if those assets were not community-owned (Knight et al., 2017). Similarly, in Mountain West communities, which are dependent upon winter recreation activities, ski areas, hotels, and other critical infrastructure are increasingly owned by large multinational corporations (Ooi et al., 2016). Non-local asset ownership influences economic leakage and minimises local economic prosperity (Dangi & Jamal, 2016) by limiting asset accumulation of locals who are increasingly forced to become seasonal or migrant industry workers (Amason, 2020; Ooi et al., 2016). As scarce economic and environmental resources are increasingly consolidated, the livelihoods of locals are increasingly threatened, questioning whether future sustainable tourism development in Mountain West communities demonstrates meaningful poverty alleviation or sustainability gains for vulnerable locals (Álvarez-Herranz & Macedo-Ruíz, 2021).

Sustainable tourism development to address property inequality

Sustainable tourism development is the coupling of sustainable development (UNWCED, 1987) and sustainable tourism (UNWTO, n.d.). It imagines sustainability as the bridge linking tourism and development, where development is "the process of creating, testing, and maintaining opportunity" in the context of tourism-based economies (Holling, 2001, p. 390). Offering clarity around sustainable tourism development remains essential, given how sustainability remains a contested concept (Giddings et al., 2002). The complexity of sustainable development is further convoluted as there is uncertainty about the origin within the academic literature, creating what Purvis et al. (2019) describe as "competing language" (p.12). Despite the various debated definitions of sustainable development, a consensus is observable in that sustainable development must consider environment, economy, and equity, frequently described as the pillars of sustainability (Becker et al., 2015; Campbell, 1996; Cottrell et al., 2007; Purvis et al., 2019; Ruiz-Ballesteros, 2011; Schively, 2008; UNWCED, 1987, 1992). The pillars of sustainability are fraught with debate over what is perceived in the literature as untenable tradeoffs amongst the pillars and development's role in antagonising these conflicts. Campbell (1996) describes this dilemma as the planner's triangle, where economic interest is frequently at odds with equity and the environment.

If sustainable tourism development is to reduce poverty by means of development that creates opportunities, then the opportunities generated must be attainable for locals. This means that locals must receive more than an income and accumulate wealth through asset accumulation. Sen (1979) describes the capability approach as a framework to evaluate human well-being using agency to pursue opportunities. Well-being is not merely achieved because the opportunity is created through development, but rather when "basic capabilities" are satisfied through development Sen (1979). For example, in the Appalachian tourist town of Gatlinburg, socioeconomically vulnerable seasonal tourism workers often utilise dilapidated mid-century motels as residences despite continued economic development, which is described as beautification of the town and prioritising increased tourist carrying capacity (Amason, 2020). The situation described in Gatlinburg relates to the findings of Narayan et al. (2004) in that persons experiencing poverty suffer the psychological awareness that their value, measured through labour, is exploited. This makes those experiencing poverty vulnerable to social, economic, and environmental harms (Narayan et al., 2004). Research indicates that sustainable tourism development can achieve asset equity, but this requires a reimagining of development that capitalises on synergies of the three pillars of sustainability (Zhou et al., 2020). This reimagining of property equity in relation to sustainable tourism development requires both resilience thinking and motivation on the part of institutional stakeholders. The capacity to generate wealth and alleviate poverty is possible through tourism, but for strong social equity outcomes, sustainable development requires community-based natural resource management and community participation (Apollo & Wengel, 2022).

Resilience of Tourism-Dependent Economies to Achieve Sustainable Tourism Development

Resilience and sustainability are sometimes used interchangeably in systems thinking. However, with clear definitions of sustainable tourism development established previously in the literature review, resilience can then be thought of as the strategy of anticipating shocks to the stability of resources to advance sustainability goals (Lew et al., 2016). Walker et al. (2004) describe resilience as being comprised of both adaptability and transformability, where adaptability is the social capability of actors to "influence the system", and transformability is the capability of a system to emerge from a foundational change in previously stable variables. Shocks are a primary concern of resilience thinking and are anticipated as either general or specific (Folke et al., 2010). General resilience is a system's ability to emerge from anticipated shocks, whereas specific resilience indicates a system's ability to emerge from a novel, unanticipated shock (Folke et al., 2010). With this understanding, the COVID-19 pandemic should be considered a novel shock, testing the specific resilience of tourism economies. The shock of the affordable housing crisis can be considered a test of the general resilience of tourism economies. As the shock began gradually, it is well understood and is increasingly anticipated by community planners in tourism destinations.

Resilience and the sharing economy

While affordable housing issues have been linked to specific in-migration motivations to rural Mountain West communities, such as second homes and access to recreation (Darling, 2005), academic literature is increasingly establishing a connection between digital platforms such as Airbnb and issues pertaining to affordable housing (Wachsmuth & Weisler, 2018). Airbnb is an industry leader in what is described as the "sharing economy" (DiNatale et al., 2018). Airbnb experienced exponential growth beginning in 2010, with 100,000 nights booked, and it grew to over 100 million nights booked in 2017 (Wachsmuth & Weisler, 2018). Multiple studies have demonstrated that the rise of the sharing economy, notably on digital platforms such as Airbnb, has resulted in significant percentages of homes being removed from housing supplies traditionally associated with residential areas. This has negatively impacted housing affordability for locals, especially socioeconomically disadvantaged populations (Gurran & Phibbs, 2017; Scheyvens & Hughes, 2019). Resilience thinking is beneficial in addressing the shock of the sharing economy to tourism destinations, as many communities in the Mountain West and elsewhere have failed to apply zoning and taxation rates like those applied to the traditional hospitality sector (DiNatale et al., 2018; Yiu, & Cheung, 2021).

Resilience thinking encourages institutional actors to develop adaptive strategies for both sharing economy participants and those displaced by increased demand for residential properties. Yiu and Cheung (2021) note that regulations should consider the potential for Airbnb to add resilience to destinations, as evidenced by the COVID-19 pandemic when large traditional hotels were mostly vacant and residential properties benefited from the flexibility of converting from short to long-term rentals. This type of flexibility is especially beneficial to sustainable tourism development resilience for communities that experience seasonal tourism. The COVID-19 pandemic also advanced the idea of remote work (Cuerdo-Vilches et al., 2021), which could further constrain affordable housing supply in remote tourist destinations such as those in the Mountain West. These new housing pressures are likely to persist, requiring sustainable tourism development actors to consider novel approaches to land-use zoning and housing density considerations (Frederick, 2022). Adopted approaches promoting destination resilience may ensure that the most vulnerable community members have access to affordable housing in the Mountain West.

Stakeholder Motivations to Address Sustainable Development Conflicts

Thus far, the literature reviewed has indicated that both conflicts over development choices and the associated impacts on the pillars of sustainability, as well as shocks introduced, either anticipated or novel, are to be expected when practising sustainable tourism development. The unanswered question is how institutional actors will address these likely occurrences. Gifford (2006) explains in the seminal paper, "A General Model of Social Dilemmas," that "each decision-maker has a set of motivations, cognitions…values" (p. 4) which influence individual actors or groups. Actor motivations may influence further conflict or cooperation (Almeida et al., 2017) when addressing challenges experienced during sustainable tourism development planning. How individual or group motivations will influence the ability of destinations to address challenges to sustainable tourism development are inherently difficult to predict or strategise for (Weaver & Lawton, 2017). However, it is understood that motivations can be either extrinsic, for public interests, or intrinsic, for self-interests (Crompton, 2010). For example, when considering conflicts between the sustainability pillars of the environment and economics, an institutional decision may be made to limit development through zoning. Ecological resources are protected by reducing land designated for development, but this may be at the expense of the equity pillar. Darling (2005) concluded that institutional land management decisions designed for ecological conservation in the Adirondacks prevented development while aiding the gentrification of Adirondack communities by increasing property values through supply side controls. Similar results have been found regarding conservation efforts and impacts on increased property values, which may motivate property owners to encourage conservation while constraining affordable housing through limiting future development (Reeves et al., 2018).

Discussion

Studies suggest that sustainable tourism development is complex and characterised by environmental, economic, social, and institutional dimensions. Research also highlights conflicts among institutional actors, or decision-makers, related to perceived tradeoffs between competing interests associated with these dimensions (UNWCED, 1987; 1992). The decision-making processes governing sustainable tourism rest on complicated value structures that influence actor motivations and public opinions (Stankey & McCool, 2004). While context-specific, the conversations addressing sustainable tourism development and affordable housing conflicts in the Mountain West are nothing new. Ewert (1996), for example, questioned how resources should be prioritised when conflicts arise between conservation and home construction. Researchers have considered the conflicts described in this chapter but in limited manners related to the context of sustainable tourism development and affordable housing in the Mountain West. The interdisciplinary nature of current research offers an avenue through which scientific foresight can be applied to these multi-dimensional issues to address what Holling (2001) describes as a tendency for social-ecological systems to meaningfully change only when disruption occurs from a catalyst.

Sustainable tourism development provides residents with economic opportunities through employment income (Knight et al., 2017). The issue resides in how those economic opportunities are distributed to residents and the ability of residents to achieve prosperity through assets. There is no panacea to this issue, but researchers can assist in providing tools to address the social equity concerns of poverty alleviation through sustainable tourism development. A beginning step could include recognising that affordability is place-based and directly related to residents" ability to access resources (Brooks, 2022). This may mean reimagining sustainability regarding how resources are made available to residents. For instance, in the United States, nearly 66 per cent of all new residential construction between 2019 and 2022 was in the form of single family detached housing (HUD, n.d.). Resilience thinking benefits this paradigm of future residential development in the Mountain West, as addressing affordability through increasing density may offer cascading sustainability benefits, including reduced carbon emission, improved resident livelihoods, and reduced land use (Frederick, 2022). Researchers can provide insight for institutional decision-makers regarding best approaches through interdisciplinary collaboration and informed understanding.

Conclusion

The multi-dimensional nature of sustainable tourism development facilitates a discussion encompassing entire academic literature journals. Determining how sustainability, resilience, motivations, and conflict can be synergised requires vast temporal and spatial fields to recognise the synergies that improve the application of sustainable tourism development frameworks. Future research could encompass many inquiries, such as how to utilise the sharing economy best to promote resilience and sustainability in tourism-based economies and what types of future residential construction best balance ecological, economic, and social equity interests. Future research could also include gathering further qualitative and quantitative data related to perceptions of affordable housing in Mountain West communities, as gaps exist regarding residents" values between conflicting resource management decisions. There is no panacea approach to solving all the issues inherent to sustainable tourism development, but continued research offers tools that can be used to promote a better future.

Management Implications

Tourism development occurs in stages proceeding from euphoric beginnings to antagonist outcomes due to complex and often subjective perspectives from the ensemble of stakeholders (Apollo & Andreychouk, 2022). Stakeholders' motivations vary but are tied to the future development of tourism destinations, regardless of spatial location, which face similar outcomes temporally. It must then be determined how to address development, which promotes rejuvenation by dispossessing decline. Regrettably, the course of action adopted has frequently resulted in the dispossession of resources from vulnerable populations in lieu of unfavourable visitation statistics. This chapter thus suggests that new indicators may improve the social equity of vulnerable host populations. These indicators should include well-being and livelihood considerations beyond gross domestic product (Stiglitz et al., 2009). Governance institutions should approach residential zoning policies with precaution and consideration for density and workforce housing accommodation. Taxation policies should recognise the housing burden contributed to local markets through absorbent vacation home rates within areas of limited housing capacity. Lastly, policymakers and destination planners in Mountain West communities may look to strengthen social safety nets for vulnerable populations who may be sensitive to the effects of ongoing, tourism-driven gentrification. Mountain tourism has the potential to be a powerful tool in addressing the goal of poverty alleviation. Destination development should seek continuous improvement strategies through community-based resource management strategies. Just transitions are the charge for governance structures and agents of power such as researchers, planners, government entities and officials, special interest, and social action organisations.

References

Almeida, J., Costa, C., & Nunes da Silva, F. (2017). A framework for conflict analysis in spatial planning for tourism. *Tourism Management Perspectives*, 24, 94–106. https://doi.org/10.1016/j.tmp.2017.07.021.

Álvarez-Herranz, A., & Macedo-Ruíz, E. (2021). An evaluation of the three pillars of sustainability in cities with high Airbnb presence: A case study of the city of Madrid. *Sustainability (Basel, Switzerland)*, 13(6), 3220. https://doi.org/10.3390/su13063220.

Amason, J. H. (2020). The making and unmaking of an Appalachian "Home": Tensions between tourism and housing development in Gatlinburg, Tennessee. *Journal of Appalachian Studies*, 26(1), 8–24. https://doi.org/10.5406/jappastud.26.1.0008.

Apollo, M., & Andreychouk, V. (2022). *Mountaineering adventure tourism and local communities: social, environmental and economics interactions.* Edward Elgar Publishing. https://doi.org/10.4337/9781802209389

Apollo, M., & Wengel, Y. (2022). *Mountaineering tourism: a critical perspective.* Routledge. https://doi.org/10.4324/9781003095323

Ashley, C., & Roe, D. (2002). Making tourism work for the poor: Strategies and challenges in southern Africa. *Development Southern Africa (Sandton, South Africa)*, 19(1), 61–82. https://doi.org/10.1080/03768350220123855.

Baron, J. S., Theobald, D. M., & Fagre, D. B. (2000). Management of land use conflicts in the United States Rocky Mountains. *Mountain Research and Development*, 20(1), 24–27. https://doi.org/10.1043/0276-4741(2000)020<0024:MOLUCI>2.0.CO; 2

Becker, C., Ewringmann, D., Faber, M., Petersen, T., & Zahrnt, A. (2015). Endangering the natural basis of life is unjust: On the status and future of the sustainability discourse. *Ethics, Policy & Environment*, 18(1), 60–67. https://doi.org/10.1080/21550085.2015.1020729.

Bell Policy Center. (2022, July 26). *Colorado housing primer.* Retrieved December 9, 2022, from https://www.bellpolicy.org/2022/07/26/colorado-housing-primer/.

Brooks, M. M. (2022). The changing landscape of affordable housing in the rural and urban United States, 1990–2016. *Rural Sociology*, 87(2), 511–546. https://doi.org/10.1111/ruso.12427.

Campbell, S. (1996). Green cities, growing cities, just cities?: Urban planning and the contradictions of sustainable development. *Journal of the American Planning Association*, 62(3), 296–312. https://doi.org/10.1080/01944369608975696.

Cottrell, S. P., Vaske, J. J., Shen, F., & Ritter, P. (2007). Resident perceptions of sustainable tourism in Chongdugou, China. *Society & Natural Resources*, 20(6), 511–525. https://doi.org/10.1080/08941920701337986.

Cottrell, S. P., Vaske, J. J., & Roemer, J. M. (2013). Resident satisfaction with sustainable tourism: The case of Frankenwald Nature Park, Germany. *Tourism Management Perspectives*, 8, 42-48. https://doi.org/10.1016/j.tmp.2013.05.005.

Crompton, T. (2010). *Common cause: The case for working with our cultural values.* Surrey, UK: WWF-UK.

Cuerdo-Vilches, T., Navas-Martín, M. Á., & Oteiza, I. (2021). Working from home: Is our housing ready? *International Journal of Environmental Research and Public Health*, 18(14), 7329. https://doi.org/10.3390/ijerph18147329.

Dangi, T. B., & Jamal, T. (2016). An integrated approach to "sustainable community-based tourism". *Sustainability* , 8(5), 475. https://doi.org/10.3390/su8050475.

Darling, E. (2005). The city in the country: Wilderness gentrification and the rent gap. *Environment and Planning A*, 37(6), 1015–1032. https://doi.org/10.1068/a37158.

DiNatale, S., Lewis, R., & Parker, R. (2018). Short-term rentals in small cities in Oregon: Impacts and regulations. *Land Use Policy*, 79, 407–423. https://doi.org/10.1016/j.landusepol.2018.08.023.

Ewert, A. W. (1996). *Natural resource management: The human dimension*. Routledge.

Folke, C., Carpenter, S. R., Walker, B., Scheffer, M., Chapin, T., & Rockström, J. (2010). Resilience thinking: Integrating resilience, adaptability and transformability. *Ecology and Society*, 15(4), 20. [online] URL http://www.ecologyandsociety.org/vol15/iss4/art20/.

Frederick, C. (2022). Economic sustainability and "missing middle housing": Associations between housing stock diversity and unemployment in mid-size U.S. cities. *Sustainability* , 14(11), 6817. https://doi.org/10.3390/su14116817.

Giddings, B., Hopwood, B., & O'Brien, G. (2002). Environment, economy and society: Fitting them together into sustainable development. *Sustainable Development* , 10 (4), 187–196. https://doi.org/10.1002/sd.199.

Gifford, R. (2006). A general model of social dilemmas. *International Journal of Ecological Economic and Statistics*, 4(6).

Gurran, N., & Phibbs, P. (2017). When tourists move in: How should urban planners respond to Airbnb? *Journal of the American Planning Association*, 83(1), 80–92. https://doi.org/10.1080/01944363.2016.1249011.

Holling, C. S. (2001). Understanding the complexity of economic, ecological, and social systems. *Ecosystems*, 4(5), 390–405. https://doi.org/10.1007/s10021-001-0101-5.

Housing and Urban Development, Department of (HUD). (n.d.). SOCDS building permits database; *Housing and Urban Development: Washington, DC, USA, 2020.* Retrieved December 9, 2022, from https://socds.huduser.gov/permits/summary.odb.

Housing and Urban Development, Department of (HUD). (2017, August 14). *Defining affordable housing*. Retrieved December 9, 2022, from https://www.huduser.gov/portal/pdredge/pdr-edge-featd-article-081417.html.

Ibanescu, B. C., Eva, M., & Gheorghiu, A. (2020). Questioning the role of tourism as an engine for resilience: The role of accessibility and economic performance. *Sustainability* , 12(14), 5527. https://doi.org/10.3390/su12145527.

IBIS World, (2022, June 23). *Tourism in the U.S. – employment statistics 2005–2028.* Retrieved December 9, 2022, from https://www.ibisworld.com/industry-statistics/employment/tourism-united-states/.

Kenney, A., (2022). *Proposition 123: Dedicate tax revenue to affordable housing*. Colorado Public Radio. Retrieved December 9, 2022, from https://www.cpr.org/2022/10/17/vg-2022-colorado-proposition-123-affordable-housing-income-tax/.

Knight, D. W. (2018). An institutional analysis of local strategies for enhancing pro-poor tourism outcomes in Cuzco, Peru. *Journal of Sustainable Tourism*, 26(4), 631–648. https://doi.org/10.1080/09669582.2017.1377720.

Knight, D. W., Cottrell, S. P., Pickering, K., Bohren, L., & Bright, A. (2017). Tourism-based development in Cusco, Peru: comparing national discourses with local realities. *Journal of Sustainable Tourism*, 25(3), 344–361. https://doi.org/10.1080/09669582.2016.1213269.

Lew, A. A., Ng, P. T., Ni, C. C., & Wu, T, C., (2016). Community sustainability and resilience: similarities, differences, and indicators. *Tourism Geographies*, 18(1), 18–27. https://doi.org/10.1080/14616688.2015.1122664.

McCool, S. F., & Bosak, K. (2015). Reframing sustainable tourism. In *Reframing Sustainable Tourism*. Springer. https://doi.org/10.1007/978-94-017-7209-9.

Narayan, D., Patel, R., Schafft, K., Rademacher, A., Koch-Schulte, S., & Rajasekhar, D. (2004). *Voices of the poor: can anyone hear us?* Oxford University Press. [PDF] Retrieved from https://documents.worldbank.org/en/publication/documents-reports/documentdetail/131441468779067441/voices-of-the-poor-can-anyone-hear-us.

Ooi, N., Mair, J., & Laing, J. (2016). The transition from seasonal worker to permanent resident: Social barriers faced within a mountain resort community. *Journal of Travel Research*, 55(2), 246–260. https://doi.org/10.1177/0047287514546226.

Purvis, B., Mao, Y., & Robinson, D. (2019). Three pillars of sustainability: In search of conceptual origins. *Sustainability Science*, 14(3), 681–695. https://doi.org/10.1007/s11625-018-0627-5.

Reeves, T., Mei, B., Bettinger, P., & Siry, J. (2018). Review of the effects of conservation easements on surrounding property values. *Journal of Forestry*, 116(6), 555–562. https://doi.org/10.1093/jofore/fvy046.

Ruiz-Ballesteros, E., (2011). Social-ecological resilience and community-based tourism. *Tourism Management (1982)*, 32(3), 655–666. https://doi.org/10.1016/j.tourman.2010.05.021.

Scheyvens, R., & Hughes, E. (2019). Can tourism help to "end poverty in all its forms everywhere"? The challenge of tourism addressing SDG1. *Journal of Sustainable Tourism*, 27(7), 1061–1079. https://doi.org/10.1080/09669582.2018.1551404.

Schively, C. (2008). Sustainable development as a policy guide: an application to affordable housing in island communities. *Environment, Development and Sustainability*, 10(6), 769–786. https://doi.org/10.1007/s10668-007-9084-5.

Scoones, I. (1998). *Sustainable rural livelihoods: A framework for analysis.* IDS Working Paper 72, Institute of Development Studies.

Sen, A. (1979, May 22). *Equality of what?* The Tanner lecture on human values. Stanford University. Retrieved from https://www.suz.uzh.ch/dam/jcr:ffffffff-df42-7cac-ffff-ffffd4ec9ff2/SEN_1.pdf.

Stankey, G. H., & McCool, S. F. (2004). Social sciences and natural resources management: An assessment of advances. In Manfredo, M. J., Vaske, J. J., Bruyere, B. L., Field, D. R., & Brown, P. J. (Eds.), *Society and natural resources: A summary of knowledge* (21–34). Jefferson, MO: Modern Litho Press.

Stiglitz, J., Sen, A., and Fitoussi, J. P., (2009). *Executive summary.* Report by the Commission on the Measurement of Economic Performance and Social Progress. Retrieved from https://ec.europa.eu/eurostat/documents/8131721/8131772/Stiglitz-Sen-Fitoussi-Commission-report.pdf.

United Nations. (1987). *Report of the World Commission on Environment and Development: Our common future.* Retrieved November 26, 2022, from https://sustainabledevelopment.un.org/content/documents/5987our-common-future.pdf.

United Nations. (1992). *Report of the World Commission on Environment and Development: Rio de Janeiro*. Retrieved November 26, 2022, from https://www.un.org/esa/dsd/agenda21/Agenda%2021.pdf.

United Nations World Tourism Organization. (n.d.). *Sustainable development*. Retrieved December 9, 2022, from https://www.unwto.org/sustainable-development.

United Nations World Tourism Organization. (2022, Nov 23). *Tourism recovery accelerates to reach 65% of pre-pandemic levels*. Retrieved December 9, 2022, from https://www.unwto.org/news/tourism-recovery-accelerates-to-reach-65-of-pre-pandemic-levels.

Wachsmuth, D., & Weisler, A. (2018). Airbnb and the rent gap: Gentrification through the sharing economy. *Environment and Planning. A*, 50(6), 1147–1170. https://doi.org/10.1177/0308518X18778038.

Walker, B., Hollin, C. S., Carpenter, S. R., & Kinzig, A. (2004). Resilience, adaptability and transformability in social–ecological systems. *Ecology and Society*, 9(2), 5–5. https://doi.org/10.5751/ES-00650-090205.

Weaver, D. B., & Lawton, L. J. (2017). A new visitation paradigm for protected areas. *Tourism Management*, 60, 140–146. https://doi.org/10.1016/j.tourman.2016.11.018.

Winter, T., & Kim, S. (2021). Exploring the relationship between tourism and poverty using the capability approach. *Journal of Sustainable Tourism*, 29(10), 1655–1673. https://doi.org/10.1080/09669582.2020.1865385.

World Travel & Tourism Council. (n.d.). *Economic impact reports*. Retrieved December 9, 2022, from https://wttc.org/research/economic-impact.

World Travel & Tourism Council. (2017, Mar 28). *Where does travel create the most jobs?* Retrieved December 9, 2022, from https://worldtraveltourismcouncil.medium.com/where-does-travel-tourism-create-the-most-jobs-4c3347670a78.

Yiu, C. Y., & Cheung, K. S. (2021). Urban zoning for sustainable tourism: A continuum of accommodation to enhance city resilience. *Sustainability*, 13(13), 7317–. https://doi.org/10.3390/su13137317.

Zhou, L., Mu, H., Wang, B., Yuan, B., & Dang, X. (2020). Evaluating urban community sustainability by integrating housing, ecosystem services, and landscape configuration. *Complexity*,1, 1–14. https://doi.org/10.1155/2020/3460962.

14 Local Food Products as an Element for Fostering Sustainable Mountain Tourism in Marginal Areas

Some Evidence from North-Western Italian Alps

Alessandro Bonadonna and Stefano Duglio

Introduction

In the European context, agriculture and tourism can play a fundamental role in the development of mountain areas. Agriculture is closely linked to the vocation of mountain territories and the production of typical and quality food products, such as dairy products. Livestock breeding, for instance, provides food for local communities, represents a source of income and employment for mountain residents and contributes to maintaining the environmental balance by preserving biodiversity and the mountain landscape. As far as tourism is concerned, this activity also relies on the natural and landscape resources of mountain areas and offers the opportunity to practice several activities and fulfil different needs and expectations (Bonadonna et al., 2017; Giachino et al., 2020; Beltramo et al., 2024). Agriculture and tourism can complement each other, providing employment and agri-food production, contributing to land conservation and income generation, and enhancing the natural and cultural heritage of the mountain territory.

An example of the interconnection between tourism and agriculture can be represented by the Bettelmatt cheese, a traditional food production in the North-West of the Piedmont Region (Italy), which is only produced in a few alpine pastures in the Antigorio and Formazza valleys. Over the past 30 years, it has reached a level of national importance, both gastronomically and in terms of tourism. Bettelmatt differs from other dairy products in the Alpine area thanks to a specific production specification that allows the final product to be endowed with specific organoleptic and sensory characteristics linked to the peculiarities of the production area (Paleari et al., 2010; Bonadonna & Duglio, 2016).

In this context, this paper aims to investigate the socio-economic implications generated by Bettelmatt cheese and its impact on the wealth of local communities from the perspective of farmers/producers, focusing on how such production can generate income and stimulate sustainable tourism in secondary mountain areas. Therefore, this contribution is organised as follows:

DOI: 10.4324/9781003392583-17

the background briefly presents both agriculture and tourism with a specific focus on mountain areas to introduce the main characteristics of this local food production (Bettelmatt). After introducing and discussing the agricultural and tourism sectors concerning the study area, the materials and methods present the methodology adopted during the analysis (semi-structured interviews). Findings focus on presenting and discussing the main results deriving from the study, while conclusions point out both limitations and new avenues of research.

Background

Agricultural and tourism sectors in mountain areas

Considering the European context, mountain agriculture represents 15% of the agricultural utilised area (Euromontana, 2023). On the other hand, however, geographical and geomorphological constraints make it difficult for mountain farmers to be competitive with other agricultural contexts in terms of productivity, which is, on average, lower by 40% compared to lowlands (Euromontana, 2023). Furthermore, European mountain agriculture is mainly characterised by small size enterprises. In Italy, for instance, around 35% of the Italian farms have less than 2 ha and around 25% between 2 and 5 ha (Rossi, 2022). Moreover, they are affected by negative conditions such as land fragmentation/availability and financial overexposure, which can represent severe barriers to starting and running mountain farms (Beltramo et al., 2018; Bonadonna et al., 2020; Duglio et al., 2023), and operate in marginal context, which has witnessed significant depopulation rates over the past decades. These are the reasons why the Italian mountain farms have declined in the last 30 years (Gretter et al., 2019). It is worth highlighting, however, that mountain farming cannot be only assessed in terms of economic implications. Even if mountain farmers are not comparable with other agricultural contexts using the profit and income lens, they play a fundamental role in preserving the landscape, the natural heritage and the cultural legacy of mountain communities (Ricciardi et al., 2021). In this perspective, mountain food products can be associated with tradition and intrinsic quality of the product itself (Duglio et al., 2022), which can be testified by the many quality system tools that have been implemented over the years, as the EU certification systems (protected denomination of origin and protected geographical indication), and labelling schemes with a particular reference to the quality scheme "mountain product" (Bonadonna et al., 2017; Bentivoglio et al., 2019).

Mountain agriculture products are usually appreciated by the agri-food market (Majkovič & Borec, 2010; Baritaux et al., 2011) and by the tourism sector, as testified by the visitors' willingness to taste and purchase local food products. In a recent study carried out in a secondary Italian mountain valley in the Alps on a sample of 507 visitors, Duglio et al. (2022) report that "gastronomy is the third most popular tourist activity, with 58.2% of tourists confirming that they had purchased local food products and 78.5% of those who had lunch in the valley saying they had tasted local recipes". These results reinforce the study of Giachino et al. (2020), which showed that "food and beverage" is one of the most important motivations for going to the mountains and is the only motivation that maintains relatively stable results in both winter and summer in terms of importance.

Thus, tourism can be a flywheel for local mountain farmers, as mountains are the world's second most popular tourism destination. In accordance with the last figures provided by FAO and UNWTO (2023), international mountain tourism is assessed to weight between 9% and 16% of international tourist arrivals worldwide, and between 195 million and 375 million tourists, and pre-pandemic analysis reported an estimated economic impact between 70 and 90 billion U.S. Dollars per year (Debarbieux et al., 2014). When considering mountain tourism, however, it is essential to underline the significant variability that can be noted among mountain tourism destinations. Mountain tourists have different needs, expectations, and motivations, which are related to soft activities (hiking, discovery, leisure, etc.) and challenging activities (climbing, alpinism, etc.) (FAO and UNWTO, 2023), and each mountain destination is not normally able to fill all these needs, but only to provide responses that are based on its own characteristics. Thus, while some mountain destinations have focused specifically on ski and mountain resorts, other destinations may focus more on offering ecotourism activities during the summer seasons based on their natural and cultural heritage, which can stimulate change and create opportunities for sustainable development (Apollo & Andreychouk, 2022).

A fil rouge for all kinds of mountain tourism destinations can be seen in the central role performed by gastronomy, which should rely on typical food products for effectiveness. Indeed, not only can food be considered a crucial element for proposing a tourism product (Ermolaev et al., 2019) but also a preeminent component of the same experience (Rachão et al., 2020). This is why an integrated coordination between local farmers and tourism operators may be a win-win strategy, especially in marginal areas (Duglio et al., 2019).

Bettelmatt cheese

Bettelmatt cheese takes its name from Alpe Bettelmatt, the last Italian alpine pasture at the foot of the Gries Pass, on the border between Ossola and Valais (Switzerland). Thanks to its geographical position, Alpe Bettelmatt was, for many centuries, a strategic point for trade routes between central Switzerland and the Po Valley (even today, the traditional "Sbrinz route" seals its former importance). Bettelmatt owes its name to the custom, dating back to the Middle Ages, of tying up wheels of cheese to distribute them on set dates to the poorest families in the community: in the Valais, the term "bettelmatt" (or bättelmatt), obtained from the union of the words "bättel" (alms) and "mattu" (meadow or pasture), means "pasture of the alms" (Rizzi, 2003). The alpine pasture's fame is due to the rich flowering of the meadow and, in particular, to the presence of Alpine or common lovage (Ligustrum Muttelina L.). The presence of this plant species gives the cheese quality characteristics that have been maintained over time (Mortarotti, 1979; Delforno, 1981; Paleari, et al., 2010).

This characteristic is shared by all the mountain pastures indicated in the Bettelmatt production specifications (Ottogalli, 2001), which were identified thanks to a study on the grazing heritage carried out in the Piedmont Region (Regione Piemonte, 2002). Between the second half of the 1990s and the beginning of the 2000s, Bettelmatt became popular thanks to Edoardo Raspelli, an expert in Italian cuisine, who spread its knowledge through the main mass media of the time (magazines and TV), reaching the homes of millions of Italians. The production of Bettelmatt follows the guidelines laid down in a production specification. The Antigorio Divedro and Formazza Mountain Community, Bettelmatt producers and alpine pasture owners have drawn up and signed a Bettelmatt production specification to define some common criteria for the production process and the product's final characteristics.

The product specification states that the name Bettelmatt identifies Ossola cheese produced in the Antigorio and Formazza valleys, with particular reference to the mountain pastures Alpe Forno and Alpe Sangiatto (Municipality of Baceno), Alpe Pojala (Municipality of Premia), Alpe Toggia/Regina, Alpe Kastel, Alpe Morasco/Bettelmatt and Alpe Vannino (Municipality of Formazza). The Toggia and Morasco alpine pastures are combined, respectively, with Alpe Regina and Alpe Bettelmatt, which share a vegetation that gives the milk characteristics of absolute value (Regione Piemonte, 2002).

Bettelmatt is a full-fat, semi-cooked cheese made from whole cow's milk from a single milking; the wheels are cylindrical with a side section of 8 cm and a diameter of about 25/35 cm and the weight can vary from 4 to 6 kilograms. The paste is yellow, compact and soft with bird's eye formation; the rind is smooth. The milk is processed immediately after milking, with the natural fermentation acidity, without pasteurisation and using starter whey. The characteristic shape is obtained using moulds with a smooth surface and salting can be dry or brine. The minimum maturing period is 60 days (Bonadonna & Duglio, 2016; Lucarno, 2022).

Furthermore, Bettelmatt cheese may not be produced below an altitude of 1,800 meters. The cattle supplying the milk must be fed with pasture grass; only trace element supplements are permitted. To ensure product recognisability, the cheeses are dated (production date) and branded with the name Bettelmatt, directly on the rind, with the EC identification mark and the alpine pasture of origin. The mark is applied by the official Bettelmatt markers only after regular evaluation and selection of the cheese wheels based on the characteristics required by the specifications (Figure 14.1). As far as marketing is concerned, each year, the producers set a minimum price for the sale of Bettelmatt at the ordinary assembly to enhance the product's value and avoid speculation.

A study conducted in the 2014/15 season highlighted the dimensions of the Bettelmatt phenomenon. At the time, data were collected from the eight Bettelmatt farms that operated in the alpine pastures specified in the specifications, except Alpe Pojala, which was not yet used. The study showed that around 500–550 dairy cows in production were taken to the alpine pastures for between 60 and 90 days. Each farm had a minimum of 40 to a maximum of 120 cows. The milk produced in the alpine pasture was (and is) used to produce Bettelmatt. The average total number of Bettelmatt wheels produced per year was just under 5,000; the average weight was 4.5–5 kg/wheel. The total quantity produced was about 23 tonnes, with an estimated production value of about EUR 520,000. This value was obtained 57% from distributors, 32% from direct sales, and 11% from retailers and/or restaurateurs. The average sales price was around 22 euro/kg (Bonadonna & Duglio, 2016).

Methodology

The study area

The area of investigation is located in the Verbano-Cusio-Ossola Province, on the northeast of the Piedmont Region, on the border with the Lombardy Region and Switzerland. This area is divided into two main parts: The Lake District and eight mountain valleys. Bettelmatt is mainly produced in two valleys, i.e., Antigorio and Formazza, which total count four municipalities: Baceno, Crodo, Premia (Antigorio Valley) and Formazza (Formazza Valley). The main territorial characteristics are reported in Table 14.1.

The area is highly characterised by agricultural activities, represented mainly by livestock breeding, as shown in Table 14.2 (ISTAT, 2022).

Figure 14.1 The Bettelmatt mark on the surface of the wheels
Source: Department of Management "Valter Cantino", Commodity Science Study Area

Table 14.1 Antigorio and Formazza Valleys

Municipality	*Surface km²*	*Inhabitants*	*Altitude (m a.s.l.)*
Baceno	68.72	871	655
Crodo	53.58	1,400	525
Formazza	130.41	446	1,280
Premia	89.17	557	800
Total	**341.88**	**3,274**	

Source: Municipalities' internet sites

Table 14.2 Farms in the Antigorio and Formazza Valleys

	Farms (no.)	*Cattle (no.)*	*Pigs (no.)*	*Horses (no.)*	*Sheep (no.)*	*Goats (no.)*	*Others (no.)*	*UAA (ha)*
Baceno	7	62	3	0	11	89	281	274.06
Crodo	17	452	12	10	21	254	66	1,162.65
Formazza	5	101	0	2	0	0	0	472.73
Premia	11	611	0	3	60	241	0	1,877.56
Total	**40**	**1,226**	**15**	**15**	**92**	**584**	**347**	**3,787**

Source: ISTAT, 2022

Note: Adult Livestock Unit (ALU)

The Antigorio and Formazza valleys sustain 40 farms, mainly devoted to breeding cattle and goats. More than 1,200 bovine heads and 600 small ruminants graze pastures at the highest altitudes during summer. When considering Premia and Baceno municipalities, the last available data report 18 farms for 673 bovine heads (59.8%) and 401 small ruminants (59.3%).

Regarding the tourism sector, Table 14.3 reports the number of accommodation facilities, beds, arrivals and the Average Length of Stay (ALS) for the year 2022 (Piedmont Region, 2022).

In the Antigorio and Formazza valleys, the tourism sector is represented by 90 accommodation facilities and 41 other rent typologies for around 4,400 beds. Considering only Premia and Baceno, the two municipalities account for about 60% of the hospitality sector and 63.7% of the beds. It is worth noting that 11 mountain huts insist on the valley (nine in Formazza and two in Antigorio) for around 473 beds, representing 10.7% of the total beds in the area. The arrivals in 2022 have been 34,719, and the foreign component weighs only 17% and this may be the main reason why the Average Length of Stay (ALS) is set to 2.21 (on average, each tourist stays 2.21 days in the area).

Table 14.3 Tourism Sector

Municipality	*Accommo- dation*	*Beds*	*Other tourism rent typology*	*Beds*	*Italian arrivals*	*Foreign arrivals*	*Total arrivals*	*ALS*
Baceno	34	2,435	22	104	7,086	1,957	9,043	2.13
Crodo	16	435	10	47	9,363	1,480	10,843	2.44
Premia	17	255	6	32	4,422	577	4,999	2.36
Formazza	23	1,114	3	12	7,861	1,973	9,834	1.91
Total	90	4,239	41	195	28,732	5,987	34,719	
Average	22.5	1,059.8	10.3	48.9	7,183	1,496.8	8,679.8	2.21

Source: Extracted from Piedmont Region, 2022

Methods

Initially, the farms belonging to the Bettelmatt Cheese Producers Association were identified. There are 8 of them, located in the Verbano-Cusio-Ossola mountain area in the municipalities of Premia (4), Crodo (1), Crevoladossola (1), Baceno (1) and Formazza (1), and they practice transhumance in the mountain pastures indicated in the production specifications. Three Bettelmatt producers operating in the mountain pastures, "Alpe Forno" and "Alpe Sangiatto", were selected as model areas for the study.

Data collection was carried out through the construction and subsequent administration of a questionnaire with a semi-structured interview aimed at analysing four main themes, i.e. general information on the farm (characteristics of the farm and its evolution over time, pastures used), general information on the territory (primary sources of wealth in the area, changes in the style and quality of life over the last 30 years, the role of the farm in the territorial context) information on the farm's production activities (main productions, social and economic benefits generated by the farm for the territory, enhancement of the cultural and landscape identity of the production area generated by the farm), the role of Bettelmatt cheese in the local context (from a historical, social, economic, environmental and tourist point of view). The interviews with the persons identified (3 producers) took place in June-July 2023 with a preliminary contact to present the study project and its objectives and then set a date for the following telephone interview. Each interview was conducted by one of the two authors and lasted, on average, about one hour. The interviews were recorded in such a way as to allow the authors to independently analyse the obtained results (Atkinson and Shaffir, 1998). The data collected from the interviews were processed individually by each author to avoid influencing each other, and, at a later stage, the results were compared and further processed, identifying the critical elements needed to reach the study's goal.

Results and Discussion

The interviews show that a typical local food product, such as Bettelmatt and related agricultural activities, can effectively sustain the territorial development of a limited geographical area, such as Alpe Devero and Alpe Crampiolo. The elements that can be considered are the creation and expansion of the Bettelmatt market, tourism promotion, cultural identity, and heritage, which are intended in their broad meaning, both cultural and natural.

Bettelmatt and economic growth

The production of Bettelmatt and the management of the related supply chain requires the collaboration of all the stakeholders, from the producers to the involvement of employees, distributors and finally, the consumers. In this way, it is possible to create direct and indirect employment opportunities contributing to local socio-economic growth. Operator One, for instance, points out how, over the last 20 years, his company has grown with the involvement of several seasonal and permanent employees. He also highlights the importance of creating related industries, recalling that:

> (…) our business generates social and economic well-being since there are employee salaries (…) plus the spin-off economy because we purchase local products such as bread, honey, local products, local wines, coffee. Thus, we generate an induced supply chain for raw materials, the entire local economy.

Furthermore, he highlights that: "Bettelmatt has generated a wealth that falls entirely on the local community. Many companies deliver to Latteria Antigoriana, and without Bettelmatt, they could not stand it. Bettelmatt is an essential element for the association (of Bettelmatt producers) and the territory system". And, "Here there have never been abandonment problems: having such a product means that all the companies have grown and continue to grow because of the economic return". The other two operators confirm these considerations. On the one hand, Operator Two reinforces this concept, as well:

> In my opinion, my company, together with the others, have contributed to developing and maintaining the valley because we create an induced supply chain, shops, etc. Moreover, we clean up the valley and give 'a big hand' to the tourism sector.

Operator Three also reinforces the concept: "Crampiolo is a small town with very few accommodation facilities. Currently, after the opening of the shop and the tasting room, I believe that we are contributing to providing a service in addition to the agricultural activity". He also adds, "In this mountain area, we also have a perspective of landscape protection and care of the landscape. At Alpe Devero and Alpe Crampiolo, we keep the pastures clean but, above all, in mowing the meadows, we give the area a certain image".

Bettelmatt and tourism

Bettelmatt generates interest from tourists and encourages their presence in the production area. The fame of this product stimulates many people to reach the places where the herds graze and the laboratories where milk is transformed. There are three cheese-making laboratories at Alpe Crampiolo, one for each farm. In accordance with previous studies (Andersson et al., 2017; Duglio et al., 2023), food and tourism are strongly interconnected, as Operator One points out:

> Bettelmatt has had an impact from a tourism point of view. There are people who come here to purchase Bettelmatt. They stay to eat, then buy a wheel of Bettelmatt or a kilo or half kilo to take back home.

Moreover, the organisation of events related to the production of Bettelmatt stimulates the tourist flow "In autumn, when we celebrate the transhumance, thousands of people come up (…). Even the simple herds grazing attract people". He also underlines the importance of Bettelmatt in terms of local catering:

> There is a flow of tourists who come just to taste it. We have it on the cutting board and people ask for it; they know, they are aware. It (Bettelmatt) has given value to all the restaurants in the area where Bettelmatt is always present. You will find typical dishes in which Bettelmatt is always included. Catering has given value to the product (Bettelmatt) and the product has given value to catering, too.

Operator Two states that the activity carried out in the Alps helps to support and increase the tourism sector: "Surely our activity generates interest: there are three sales outlets in Crampiolo so certainly yes, to us or others, tourists can go to buy something of our productions (including Bettelmatt)". Furthermore, he underlines that the tourism flow is also fuelled by the territory and the pastures from which Bettelmatt and other mountain pasture cheeses are obtained. In fact, "The area is beautiful and even the pastures are beautiful, among the best here in the northern part of the Piedmont Region".

Operator Three further strengthens the role of Bettelmatt from a tourist point of view by highlighting the curiosity that this cheese generates in tourists ("We receive visits directly to the mountain pastures when we produce Bettelmatt. (…). Attracting tourists 2–3 hours from where they leave the car is not so easy"). Furthermore, Bettelmatt is used in the most renowned restaurants: "It is a much sought-after product in haute cuisine and, therefore, some of the best chefs use this product. In doing so, they advertise both for their preparations and the territory".

Bettelmatt and local economy

The interviewees also highlight the role of Bettelmatt in supporting and improving agricultural activities, defining it as a valuable element for local territorial development. Operator One repeatedly highlights the importance of Bettelmatt in generating direct and indirect wealth: "If you sell it (Bettelmatt) at 24 euros per kilo, you can sell the others (kinds of cheese) at 16/18 euros", which is useful for fuelling modernisation processes and company's innovation by improving the quality requirements of the final product because "the introduction of technology today allows us to make an excellent standardised product, and avoid disasters (in term of health diseases)". This reflection supports the consideration of technology's central role in the food industry (Singh & Bagade, 2022). This concept is further strengthened by Operator Two, who argues that:

> having a mountain pasture where Bettelmatt is produced gives you the opportunity to invest more money. We pay attention to production because you always have to do your best to guarantee the best result. The greatest value is reinvested for the environment and for the technology needed to make the product standardised.

Operator Three highlights the role of Bettelmatt as a valuable tool for generating profit: "Bettelmatt primarily contributes to the maintenance of agricultural activities in the mountain pastures. Because of its quality the market perceives it as a high-quality value product. Therefore, the price of this product also has a good response". In this way, it encourages the young to live in the mountains ("It makes even the youngest kids want to keep the activities") as well as to invest in the area:

> Not only has Bettelmatt increased the income for the agricultural farms, but also the values of mountain pastures. Leases, in the past, were quite low (mountain pastures are owned by local municipalities), while today they are seen as a resource for the municipality (they have substantially increased), because the rents can be reinvested to improve services in the area.

Bettelmatt and traditions

Bettelmatt also appears to play a role in maintaining local cultural identity and heritage.

According to Operator One

> when you talk about Bettelmatt, gastronomic experts are able to identify the territory through the cheese. In this regard, I can tell an anecdote: I'm a sommelier, and some time ago I had to go to Turin to collect an award together with …. (family name), the first guy to produce wine in Ossola. After the award ceremony, …. (family name) and I went to a restaurant where we talked with the owners. On that occasion, it emerged that they knew little or nothing about the Ossola Valleys products but, without any shadow of a doubt, they were aware of the existence of Bettelmatt and recognised its various qualities.

Operator Two shares the same feelings regarding the perception of Bettelmatt and its fame:

> Many times people do not know where Ossola is, but they do know where Bettelmatt is produced. Many times, you hear people say "ah, it's

> here where Bettelmatt is made!" and customers are also attracted by the name because they know it.

Operator Three underlines how cultural identity is integrated into the same name of the farm, which recalls the mountain and the dairy cows: "A name that refers to the kind of animal. The name was chosen for its reference for the territory". Furthermore, it points out the curiosity of the Bettelmatt consumer who, after knowing the product, wishes to have more information on the area of origin: "Many tourists have known the mountain pasture after knowing the Bettelmatt. This is probably because the consumer, appreciating the product, is curious to see where it is made". Operator One also highlights the role of Bettelmatt in maintaining the tradition of the Walser culture in the area

> There is a historical connection between Alpe Sangiatto and Alpe Forno with the Walser as all the mountain pastures in the area were inhabited by the Walser, even if they were mainly present in Val Formazza. This cheese has a centuries-old history with a production process that has remained constant over time.

Furthermore, Operator One underlines the role of Bettelmatt in terms of recovery and protection of the natural heritage: "Bettelmatt has stimulated the recovery of Alpe Pojala. A company is using it to produce Bettelmatt ("load it"). (…) It certainly allows the recovery of the territory from a landscape point of view". Operator Three shares the same opinion:

> Mountain pastures, especially here in Alpe … (name), are difficult to manage because there is a very small road and you can reach it only with a small tractor. There are logistical difficulties and production would have stopped if there hadn't been a global interest in the type of product (Bettelmatt). We have seen uncomfortable mountain pastures that have been abandoned.

He also highlights the role of Bettelmatt in keeping the activities in a good state

> In this sense, Bettelmatt was precisely 'the architect' of keeping alpine pastures active today, otherwise they would be closed. Alpe … (name) is an example: (…) it is an uncomfortable pasture and it can only be reached on foot. It doesn't have a road and therefore, you know, managing a herd, managing a production, then managing the products is not that simple.

Bettelmatt and socio-economic conditions

Finally, Bettelmatt is indicated as a tool for improving local socio-economic conditions. Operator One points out that "Bettelmatt has substantially contributed to the quality of life in the last 20 years, since, as a leading product, it has also driven all the other products". The local residents, however, do not always seem to realise it: "The natives do not realise it". Agricultural activities have been maintained over time and have grown thanks to Bettelmatt. The definition of the brand and the establishment of the association made it possible to give value to dairy production:

> Cheese was finally paid for its value. It didn't happen before". Furthermore, it allowed the creation of a stable dairy sector: "Many companies deliver to the Antigoriana Dairy and without Bettelmatt they could not survive. Bettelmatt is an essential element for the local system.

Operator Two highlights that:

> Bettelmatt is a niche cheese, elite and known throughout northern Italy. It is also exported abroad and this has made it possible to enhance a product and therefore to bring more well-being. If you sell it at 22/23 euros per kilo, instead of at 15, more value remains in the area. Let's say that Bettelmatt also drives other products. Quantifying this phenomenon is difficult but it can give a widespread benefit". In essence, the concept can be summarised with "Bettelmatt acts as a driving force and generates well-being in the local area".

Finally, Operator Three specifies that, compared to 20–30 years ago,

> There has been a technological innovation and also a continuous generational changeover and, therefore, the children have also brought innovations in terms of mechanisation and technology to further improve production activities. In the transformation of milk into cheese, too.

In the past "We produced milk and our cheese, we aged it and brought it to the valley floor for selling it to our customers during the winter season". Currently, "A tourist who goes up to Alpe …. (name) has one more opportunity: he can go to eat polenta in one of the three restaurants or have a tasting of cheeses, yoghourt, dairy products directly in one of the company outlets".

Conclusion

Local development in a rural area passes through several tools, such as local development programs capable of stimulating and supporting the creation of infrastructure, economic diversification, sustainable agriculture and multifunctionality, tourism, job offers, conservation of culture and heritage, promotion of the creation of a more cohesive community environment. In this context, a local element that can integrate the fundamental pillars of local development and contribute to reaching these goals is the local food products. Based on the evidence from this study, Bettelmatt seems to be a valid example to confirm that a food product, entirely produced in a small and delimited area, can effectively support the local economy and stimulate its growth. Bettelmatt seems to respond positively to the maintenance of agricultural activities in marginal areas and in harsh environments (e.g. pastures and high-altitude dairies), the protection of cultural identity (e.g. traditions, transhumance), the conservation of the local heritage (i.e. cleaning and maintenance of paths and pastures), the tourism promotion (i.e. a tourist flow fed by the supply chain), up to the generation of wealth that falls entirely on the local area (e.g. modernisation of production activities, recovery of farmhouses, improvement of paths).

Furthermore, implementing and/or strengthening a local production system, such as the Alpe Devero/Bettelmatt connection, has potential positive consequences for operators and stakeholders that must be considered at the community level. On the one hand, one of the main benefits is the creation of economic opportunities for the local community. Indeed, the opening of new businesses or the strengthening of existing ones, as in the case of Alpe Devero, increases employment in the area, contributing to economic growth and improving residents' quality of life. On the other hand, local actors must be aware of the challenges that may emerge during this process. In this sense, the stakeholders in the community and in the supply chain, i.e., the local population, local administrators and institutions at different levels, and business partners, must create positive relationships through careful management of local dynamics to avoid conflicts and misunderstandings. In particular, local authorities must be aware that Bettelmatt generates wealth for the entire region through the "triple bottom line" i.e. economic prosperity, social responsibility and environmental protection.

Additionally, the authorities themselves must be aware of the weight of their decisions since, based on the choices made, they can influence the modernisation of farms, the involvement of younger generations in local entrepreneurial activities, the improvement of services in mountain areas and the increase of local tourist flows. These results can only be achieved through constant collaboration between all the local players. In this sense, in the recent history of Bettelmatt, a fruitful collaboration has already taken place considering the efforts made by all stakeholders to define a production specification and related quality system and the promotional activities developed to enhance local production and the natural, cultural and historical heritage. The results obtained from the study seem to confirm what was hypothesised in the research project. However, some limitations are evident: first, the only involvement of Alpe Devero and Alpe Crampiolo operators (Alpe Forno and Alpe Sangiatto, the Bettelmatt pastures concerned) and, second, a partial analysis of the phenomenon due to the only primary stakeholders' viewpoint. In this sense, therefore, future research must be oriented towards the involvement of other stakeholders (e.g. local authorities, branders, tourists, associations) and the collection and analysis of the related perspectives to have a complete picture of the role of Bettelmatt on local development.

Management Implications

The creation of a production system shared by farmers in the mountain environment, with the participation of other local actors, can contribute to the territory's economic growth and sustainable development. An environment of collaboration and mutual support between agricultural enterprises, public bodies and economic operators makes it possible to reach critical mass for easier access to financial resources, allows for more targeted and widespread promotion of the area, enables the sharing of good practices and optimises the allocation of available resources. The creation of a network of local stakeholders, therefore, can be interpreted as a tool that more easily leads to the implementation of integrated quality systems aimed at safeguarding the territory and biodiversity, improving the quality of life in challenging environments, and generating income for the younger generations, who are incentivised to monitor and maintain the territory. In short, creating a reliable and cohesive network of local stakeholders can lead to achieving objectives to enhance the mountain territory and its resources through the maintenance of the local tangible and intangible heritage. In conclusion, developing a production system in the local area offers numerous economic opportunities but requires careful management of relationships, environmental and social implications and available resources. Land managers must be prepared to meet these challenges by adopting strategies and making decisions that ensure the local community's social, economic and environmental success.

Acknowledgements: Heartfelt thanks are due to the Producers of Bettelmatt for their kind collaboration.

References

Agriculture and rural development. (s.d.). Euromontana. Available on 25 July 2023, at https://www.euromontana.org/en/working-themes/agriculture-and-rural-development/.

Andersson, T. D., Mossberg, L., & Therkelsen, A. (2017). Food and tourism synergies: Perspectives on consumption, production and destination development. *Scandinavian Journal of Hospitality and Tourism*, 17(1), 1–8. https://doi.org/10.1080/15022250.2016.1275290.

Apollo, M., & Andreychouk, V. (2022). *Mountaineering Adventure Tourism and Local Communities.* Edward Elgar Publishing. https://doi.org/10.4337/9781802209389.

Atkinson, A. A., & Shaffir, W. (1998). Standards for Field Research in Management Accounting. *Journal of management accounting research.* Available at SSRN: https://ssrn.com/abstract=137268.

Baritaux, V., Tebby, C., & Revoredo-Giha, C. (2011). How well do food retailers know their customers? The case of mountain food products in Europe. *Journal on Chain and Network Science*, 11(3), 223–234. https://doi.org/10.3920/JCNS2011.x198.

Beltramo, R., Bonadonna, A., Duglio, S., Peira, G., & Vesce, E. (2024). Local food heritage in a mountain tourism destination: Evidence from the Alagna Walser Green Paradise project. *British Food Journal, 126(1), 309-326.* https://doi.org/10.1108/BFJ-12-2022-1136.

Beltramo, R., Rostagno, A., & Bonadonna, A. (2018). Land Consolidation Associations and the Management of Territories in Harsh Italian Environments: A Review. *Resources*, 7(1), 19. https://doi.org/10.3390/resources7010019.

Bentivoglio, D., Savini, S., Finco, A., Bucci, G., & Boselli, E. (2019). Quality and origin of mountain food products: The new European label as a strategy for sustainable development. *Journal of Mountain Science*, 16(2), 428–440. https://doi.org/10.1007/s11629-018-4962-x.

Bonadonna, A., & Duglio, S. (2016). A Mountain Niche Production: the case of Bettelmatt cheese in the Antigorio and Formazza Valleys (Piedmont – Italy). *Quality – Access to Success*, 17 (150), 80–86.

Bonadonna, A., Peira, G., Giachino, C., & Molinaro, L. (2017). Traditional Cheese Production and an EU Labeling Scheme: The Alpine Cheese Producers' Opinion. *Agriculture*, 7(8), 8. https://doi.org/10.3390/agriculture7080065.

Bonadonna, A., Rostagno, A., & Beltramo, R. (2020). Improving the Landscape and Tourism in Marginal Areas: The Case of Land Consolidation Associations in the North-West of Italy. *Land*, 9(6), 6. https://doi.org/10.3390/land9060175.

Debarbieux, B., Oiry Varacca, M., Rudaz, G., Maselli, D., Kohler, T., & Jurek, M. (2014). *Tourism in mountain regions: Hopes, fears and realities.* Department of Geography and Environment, University of Geneva; Centre for Development and Environment, University of Bern.

Delforno, G. (1981). *I formaggi tipici del Piemonte e della Valle d'Aosta*. Eda Edition.

Duglio, S., Bonadonna, A., & Letey, M. (2022). The Contribution of Local Food Products in Fostering Tourism for Marginal Mountain Areas: An Exploratory Study on Northwestern Italian Alps. *Mountain Research and Development*, 42(1), R1–R10. https://doi.org/10.1659/MRD-JOURNAL-D-20-00075.1.

Duglio, S., Bonadonna, A., Letey, M., Peira, G., Zavattaro, L., & Lombardi, G. (2019). Tourism Development in Inner Mountain Areas – The Local Stakeholders' Point of View through a Mixed Method Approach. *Sustainability*, 11(21), 5997. https://doi.org/10.3390/su11215997.

Duglio, S., Salotti, G., & Mascadri, G. (2023). Conditions for Operating in Marginal Mountain Areas: The Local Farmer's Perspective. *Societies*, 13(5), 107. https://doi.org/10.3390/soc13050107.

Ermolaev, V. A., Yashalova, N. N., & Ruban, D. A. (2019). Cheese as a Tourism Resource in Russia: The First Report and Relevance to Sustainability. *Sustainability*, 11(19), 5520. https://doi.org/10.3390/su11195520.

Food and Agriculture Organization of the United Nations and World Tourism Organization. (2023). *Understanding and Quantifying Mountain Tourism*. FAO/UNWTO. https://www.e-unwto.org/doi/book/10.18111/9789284424023.

Giachino, C., Truant, E., & Bonadonna, A. (2020). Mountain tourism and motivation: Millennial students' seasonal preferences. *Current Issues in Tourism*, 23(19), 2461–2475. https://doi.org/10.1080/13683500.2019.1653831.

Gretter, A., Torre, C. D., Maino, F., & Omizzolo, A. (2019). New Farming as an Example of Social Innovation Responding to Challenges of Inner Mountain Areas of Italian Alps. *Journal of Alpine Research, 107–2*. https://doi.org/10.4000/rga.6106.

Lucarno, G. (2022). Il formaggio bettelmatt: rapporti tra cibo e territorio nella tradizione walser della Valle Antigorio-Formazza (Piemonte). In C. Spadaro, A. Toldo & E. Dansero *(Eds.), Geografia e cibo: ricerche, riflessioni e discipline a confronto (55-62)*. Società Studi Geografici.

Majkovič, D., & Borec, A. (2010). Are consumers in Slovenia concerned about the mountain quality food? *Revija Za Geografijo*, 5(1), 115–123. https://dk.um.si/IzpisGradiva.php?id=70045.

Mortarotti, R. (1979). *I Walser nella Val d'Ossola*. Libreria Giovannacci Edition.

Ottogalli, G. (2001). *Atlante dei formaggi*. Hoepli Edition.

Paleari, M. A., Beretta, G., & Vanoni, L. (2010). An old traditional mountain cheese: The «Bettelmatt®»: microbiological and chemical characteristics. *Industrie Alimentari*, 49 (508), 23–31.

Rachão, S., Breda, Z., Fernandes, C., & Joukes, V. (2020). Cocreation of tourism experiences: Are food-related activities being explored? *British Food Journal*, 122(3), 910–928. https://doi.org/10.1108/BFJ-10-2019-0769.

Regione Piemonte (2002). *Il formaggio Ossolano. Uno studio per la caratterizzazione del territorio, dei sistemi produttivi zootecnici e dei formaggi*. Regione Piemonte – INTERREG II Italia Confederazione Elvetica.

Ricciardi, V., Mehrabi, Z., Wittman, H., James, D., & Ramankutty, N. (2021). Higher yields and more biodiversity on smaller farms. *Nature Sustainability*, 4(7), Articolo 7. https://doi.org/10.1038/s41893-021-00699-2.

Rizzi, E. (2003). Bettelmatt: l'oro dei Walser. *Pumatterblatt*, 10(45), 3–15.

Rossi, Rachele. (2022). *Small farms' role in the EU food system*. European Parliament. https://www.europarl.europa.eu/RegData/etudes/BRIE/2022/733630/EPRS_BRI(2022)733630_EN.pdf.

Singh, N., & Bagade, P. (2022). Quality, Food Safety and Role of Technology in Food Industry. In N. Kumar, A. Panghal & M. K. Garg *(Eds.), Thermal Food Engineering Operations* (415–454). John Wiley & Sons, Ltd. https://doi.org/10.1002/9781119776437.ch14.

Tourism in mountain regions: Hopes, fears and realities. (2014). Swiss Agency for Development and Cooperation. https://archive-ouverte.unige.ch/unige:159819.

15 Regional Economic Perspective on the Transformation Of Glacier Tourism in the Era of Climate Change

Emmanuel Salim and Leïla Kebir

Introduction

As objects, glaciers have been of interest for centuries, have contributed to the development of mountaineering tourism (Apollo & Wengel, 2021), and have been used for tourist activities in Arctic or mountainous areas (Welling et al., 2015). In the European Alps, glacier tourism is now important to many local tourist economies. Major tourist destinations in France, Switzerland and Austria attract millions of visitors each year thanks to glaciers (Salim, Gauchon, et al., 2021). This development is also present in Canada (Lemieux et al., 2018), New Zealand (Purdie, 2013), or China (Wang et al., 2010).

However, climate change is now a major driver of change in glacier tourism. Glaciers worldwide are retreating rapidly as temperatures rise (Roe et al., 2017). This has a number of implications for glacier tourism: access to the glacier or to itineraries around the glacier becomes more difficult and sometimes impossible; activities become more dangerous; infrastructure is destabilised; and the overall attractiveness of the site may be diminished (Salim, Ravanel, et al., 2021). At the same time, visitor behaviour is changing in terms of their perception of the glacial landscape. One example is the development of Last Chance Tourism (LCT), defined from a tourist perspective as the willingness to see a feature before it disappears (Lemelin et al., 2010). LCT is nowadays an ongoing dynamic in glacier tourism sites (Lemieux et al., 2018; Salim & Ravanel, 2023).

Previous work on glacier tourism and climate change has examined geomorphological changes and stakeholder perceptions and adaptation (Mayer & Abegg, 2022; Purdie, 2013; Salim, Ravanel, et al., 2021), visitors behaviours and reactions to the changing landscape (Groulx et al., 2016; Moreau, 2010; Weber et al., 2019). However, few studies have examined how climate change affects the economic system of glacier tourism and its implications for territorial development. Therefore, this chapter aims to present the processes that led to the construction of glaciers as cultural and economic resources and how these processes are now being questioned and redirected by the consequences of climate change from a territorial economic perspective (Colletis-Wahl et al., 2008).

DOI: 10.4324/9781003392583-18

Literature review

Regional economy and production system

From a regional economic perspective, a resource is defined as a relational process between a production system and an "object" (Figure 15.1; Kebir & Crevoisier, 2007). Accordingly, a resource does not exist by itself but is revealed and set up by actors who constitute the production system (De Gregori, 1987). In this perspective, a territorial resource is considered as a territorial characteristic built by actors in a development perspective (Gumuchian & Pecqueur, 2007). The territorial resource is, therefore embedded in time and space. The resource evolves in a relational approach with respect to its production system, defined as all the actors involved in the production of a good or service (Kebir, 2016). The production system identifies an object, sets a production intention, and implements the means to achieve it.

The conceptual framework presented in Figure 15.1 is based on the idea that there is a relational process between an object, a production system, and a consumption system. According to Kebir & Crevoisier (2007), the object (artefact, knowledge or, for our purposes, a glacier) is identified by a set of actors and their production means. The cognitive process of identification implies that actors change their representation of the object in order to imagine a possible use and valuation of it. In this perspective, a resource is defined by the relational link between an object and a production system. A resource exists as soon as a production intention is projected onto an object (Kebir, 2016).

Although the focus of the analysis of value creation has initially been on the relationship between the object and the production system, it seems that a third system is important and is described by Kebir (2016) as the consumption system. The consumption system relies on the idea of the market and its involvement in the construction of value. This system incorporates the consumer and his or her logic of satisfaction, the distributor of the good or service (located between the producer and the consumer), the actors who generate demand or evaluate it, or the media.

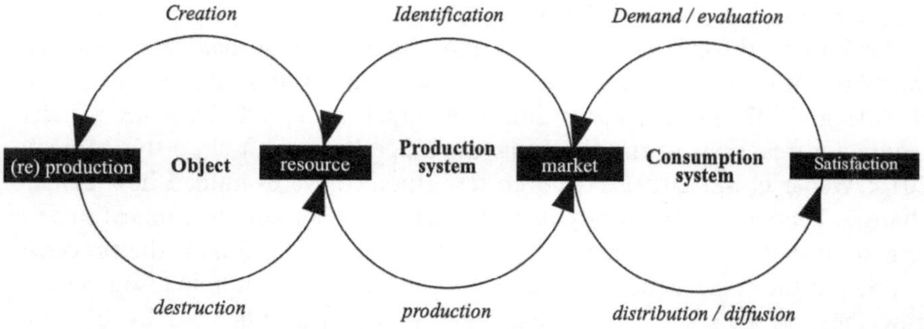

Figure 15.1 Conceptualisation of the economic system of valuation (modified from Jeannerat & Kebir, 2016)

In this perspective, and according to Kebir (2016), the resource is understood here as the relational process between an object, a production system, and a consumption system. The analysis of the resource from this perspective has been developed in the world of watchmaking and industry, or in the field of energy production (Jeannerat, 2021), But little interest has been shown in research on creating value from a natural object through a tourism approach, and even less in the case of an object directly affected by the consequences of climate change.

Glacier tourism transformation

Glacier tourism can be seen here as the result of a relational process between a glacier as an object (the glacier itself or its representation), a system of actors producing its tourism at sites, and the visitors to these sites with their expectations. In this system, the glacier and its immediate environment are largely subject to intense geomorphological processes due to climate change. These processes are characterised by rapid glacier retreat, land movements due to post-glacial decompression, increased collapse due to permafrost warming and progressive detrital glacier cover (IPCC, 2022).

These processes have direct impacts on glacier tourism. Access to the glacier is becoming increasingly difficult or impossible, as in New Zealand (Purdie, 2013) and Canada (Lemieux et al., 2018). The practices, especially mountaineering, are more difficult or even more dangerous, for example, in France (Mourey et al., 2019). The degradation of the permafrost in areas that are sometimes equipped with tourism-related infrastructure is leading to the destabilisation of the latter (Duvillard et al., 2015). As noted in the European Alps, these consequences lead to site management issues, safety and even attractiveness (Salim et al., 2021).

Climate change is transforming the physical conditions for glacier tourism activities and has an impact on the production system (ending of certain activities, etc.). The ongoing disappearance of glaciers changes the landscape, which poses a problem for the attractiveness of sites. For example, studies conducted at Yulong Snow Mountain in China (Yuan et al., 2006) and Glacier Country in New Zealand (Stewart et al., 2016) show that a significant proportion of visitors say they would not come to these areas if there were no glaciers to see. Conversely, the disappearance of the glaciers leads to the existence of LCT, which visitors wish to see before it is gone. This motivation has been observed in the European Alps (Salim et al., 2023), Iceland (Welling et al., 2020), New Zealand (Stewart et al., 2016) and Canada (Lemieux et al., 2018).

Therefore, this chapter aims to answer the question: how do the consequences of climate change lead to the transformation of production and consumption systems, as implemented in the case of glacier tourism?

Investigating the value production of glaciers in the European Alps

In order to answer the research question proposed above, 22 semi-structured interviews were conducted between 2019 and 2022 with stakeholders of 4 glacier tourism sites located in France, Switzerland and Austria. The sites in question (Fig. 15.2) were selected on the basis of the inventory carried out by Salim, Gauchon, et al. (2021) in order to represent sites at modest altitudes (below 2000 m a.s.l.) where the consequences of climate change might be more intense, and sites of importance with many visitors and infrastructures. In addition, the selected sites are built around iconic glaciers for each country.

The semi-structured interviews were conducted with key stakeholders at each site, including public operators (tourist office, council representative, park manager, N=8), private companies exploiting the site (N=9) and individual entrepreneurs such as mountain guides, mountain leaders or consultants (N=5). Interviews were conducted on-site from September to November 2019 (7), March to June 2021 (7), and October 2022 (3). The remaining five interviews were conducted online in winter 2020 due to the COVID-19 pandemic. Interviewees were selected using purposive sampling (Patton, 2002).

The interview guide varied according to the specificity of each site, but the common themes were (i) the role of the interviewee, (ii) the location of the site in relation to the overall destination, (iii) the perceived impacts of climate change, (iv) adaptation and transformation of the site, and (v) consequences of

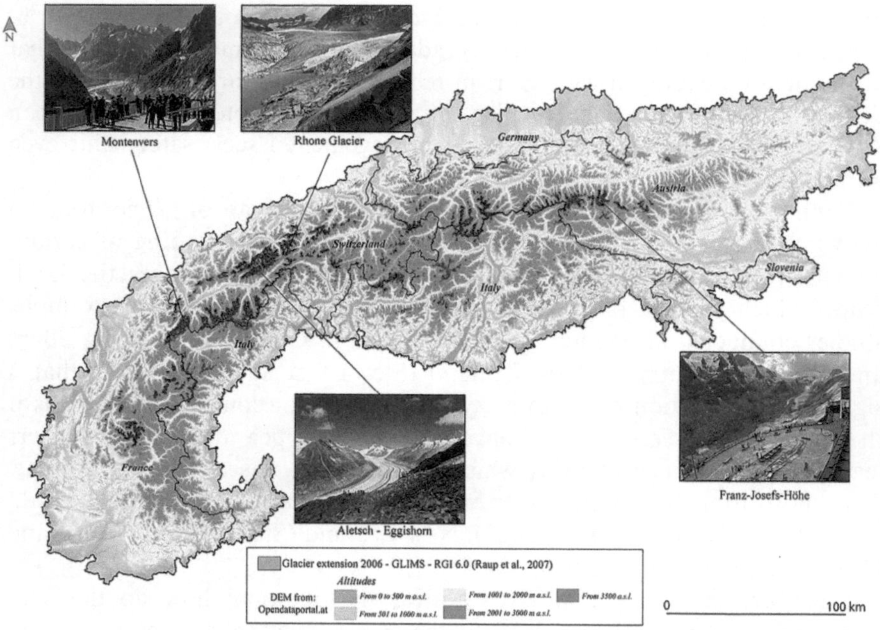

Figure 15.2 Location of the sites include in the analysis – Picture from the author, summer 2019 and 2020 (Franz-Josefs-Höhe)

the transformations and vision for the future. All interviews lasted between 1 and 2 hours and were recorded and transcribed with the participant's consent.

From Traditional to Post-Glacial Valuation

The traditional valuation of glaciers

Glaciers have no economic value in themselves. It is even clear that, historically, glaciers were primarily objects of repulsion because of the problems they caused for local economies. In the mid-18th century (for the Mer de Glace) and in the 19th century (for the other glaciers), the glaciers studied became tourist resources and sources of added value. Windham and Pococke, both Englishmen, can be considered the first people to identify the Mer de Glace as a resource when they gave it a name during their visit in 1741 and published their voyage, describing the sublime nature of the glacier (Joutard, 1986). The development of glacier tourism as it stands now with large infrastructures emerged in the middle of the 20th century. After the first stages of organisation of local actors to form a tourist offer (through the construction of hotels, excursion guides, publication of guidebooks, and access infrastructure to the glaciers), which were described in particular for the Mer de Glace by Salim, Mabboux, et al. (2021), the production and consumption system where built on the proposition of activities and experiences for tourists led by local mountain guides, transport providers, etc.

The interviewees all described the importance of the aesthetic value of the glacier and the surrounding landscape in tourism development. This aesthetic aspect encouraged the construction of the Hotel Belvédère near the Rhone Glacier in the 19th century. The Mer de Glace and the Rhone Glacier operators then developed means of providing access to the glacier by building ice caves and charging admission. The two ice cave cases studied show differences and similarities in their production systems. In the case of the Rhone Glacier, the entire site (access road, ice cave, shop and restaurant) is owned by a single family. In the case of the Mer de Glace, the ice cave and its construction are owned by a family company, but elements such as access infrastructure, safety management and advertising are produced by another private company in response to a public-private partnership. In the four cases studied, the production system is relatively limited. It includes the infrastructure for the construction of the ice caves, where they exist; the hotels, which are often located outside the site (the Eggishorn and the Franz-Josefs-Hohe have no hotel on site); and the souvenir shops, which are always present and managed by the companies in charge of access to the site.

The consumption system is much more complex. In all cases, it includes the infrastructure that provides access to, or views of, the glacier. In the case of the Aletsch and Mer de Glace glaciers, these are mechanical lifts (cable cars and cogwheel railways) managed locally by private companies but financed nationally by public bodies for their construction. The case of the Rhone is different, as the access road was primarily designed as a transit route between the cantons of Valais and Uri via the Furka Pass. The case of the Franz-Josefs-Hohe is similar to the first two, except that it is a "panoramic" route managed by a private operator. In all cases, the promotion of the sites is traditionally done through the communication of the private operators, supported by the destination management organisations. For the interviewees, the traditional motivation for visitors is to see the glacier, the sublime, the landscape, and the aesthetic experience. In summary, the glacier sites studied are characterised by a relatively limited production system, consisting of accommodation, souvenir shops and sometimes infrastructure allowing "intra-glacial" scenography. The consumption system is more complex and includes all the access and promotion infrastructures, as well as the visitors' motivations.

Impacts of climate change on the valuation system

Climate change directly affects the object of the system (the glacier) and induces changes in the value system. One notable change in the production system is the increasing difficulty of digging ice caves. In the case of the Rhone glacier, this is due to the retreat and loss of thickness of the glacier. In the case of the Mer de Glace, it is due to the reduction in the flow velocity of the glacier. As a result, the ice cave cannot be built in the same place for several years in a row, forcing a lengthening of the access, which is sometimes prohibitive for tourists. Respondents now perceive the retreat of the glacier as a threat to the site's aesthetics and its attractiveness. This leads to a reinforcement of the means developed by the production system, notably through the construction of glacier interpretation centres where the glacier, its dynamics and its history are presented. In cases where this has been developed (Mer de Glace and Pasterze), this new offer is integrated into the site's experience at no extra cost.

However, it is in the consumption system that the changes associated with glacier retreat are most significant. Access to the glaciers is one of the main concerns of the operators, particularly in the case of Mer de Glace, Aletsch and Pasterze, with the exception of the Rhône glacier, which is very accessible. In all three cases, the glacier retreat is forcing operators to create new access infrastructures, ranging from steps and footbridges to new gondola projects. Here, the degradation of the object (the glacier) leads directly to an increase in the resources allocated by the consumption system. Questions of promotion are also raised. In the case of the Mer de Glace case, promotion focused on the glacier and its aesthetics. Nowadays, it focuses on the heritage dimension of the site, which is linked to the history of glaciology and mountaineering. At the same time, a new actor has entered the consumption system: the national, regional and local media. For example, glacial retreat is the most frequently cited example of climate change in the media, with the Rhone and Aletsch glaciers in Switzerland and the Mer de Glace in France among the most frequently cited examples. The consumption system is thus partly escaping the traditional actors as a result of the degradation of the object. Finally, and possibly in response to these changes, visitors' motivations seem to have incorporated the ongoing degradation of the object through the emergence of last-chance tourism. At the sites studied, Salim et al. (2023) show that more than half of visitors come to see the glacier with the idea of seeing it before it disappears and with the idea of better understanding climate change issues.

As a result, the destruction of the object leads to an intensification of the means allocated by the production system and a transformation of the consumption system through an intensification of access, a change in the type of promotion aimed at transforming the object itself within the system, the appearance or reinforcement of an external actor that intensifies the promotion of the glacier object, and a transformation of the motivations of visitors.

The future of glacier valuation for tourism in the climate change era

The resource can follow different dynamics, ranging from renewable growth, erosion or depletion, driven by the production system, to scarcity, driven by the object (Kebir, 2006). In the case of glacier tourism, the retreat of the glaciers leads to a dynamic of scarcity of the resources. For Kebir (2006), the dynamic of scarcity ultimately leads to the disappearance of the object and, therefore, of the resource. In this sense, either the object is replaced or the activity is abandoned. The interviews show, on the one hand, that resource scarcity (glacier shrinkage) leads operators to develop access infrastructures that have no impact on the object, but in a sense "slow down" the scarcity. On the other hand, some strategies, such as the development of interpretation centres (Nesur et al., 2022), lead the production system to generate value not from the glacier itself as an object, but from its artefacts related to knowledge (glacier and climate science) or history (mountaineering and tourism development). In this sense, some strategies lead operators to replace the glacier as an object with another object.

Salim (2023) has tried to imagine what the future of glacier tourism could be, based on current developments and proposed different paths related to geotourism, dark tourism, virtual reality, or also hyper-tourism. From a pro-poor tourism perspective, changes in the valuation system of glacier tourism due to its future retreat could have implications for equity and justice. On the one hand, the development of geo-tourism, defined as a nature-based tourism activity based on a particular element of the geosystem (Newsome & Dowling, 2010), could provide activity for different entrepreneurs and access to this activity at a moderate cost. It implies a transformation in the production system by means of a more diversified offer and by placing the glacier as a resource, not only based on the traditional aesthetic value but also on the scientific one (Bussard & Reynard, 2022). On the other hand, future development based on the perspective of hyper-tourism, defined as the tourism of excess, involving high-cost infrastructure and luxury positioning (Bourdeau,

2018), has the potential to reduce access to activities to the high number and developing tourism activities based on luxury and carbon intensive travel. In all cases, the future development proposed by Salim leads the production system to replace the glacier as an object with scientific knowledge in the case of geo-tourism, with memory and heritage in the case of dark tourism, and with "virtual glaciers" in the case of the Morterasch.

Conclusion

Using four alpine glacier tourism sites as examples, this chapter has attempted to provide an overview of how the effects of climate change are currently transforming value production systems around glaciers. In particular, it shows that in order to better understand the economic implications of climate change for tourism, it is consistent to read these changes through the lens of regional economics and the theory of the economic system of valuation. In the case of glacier tourism, the transformation of production and consumption systems occurs as a result of physical changes in the environment but also through the appropriation of these changes by media originally external to the system under consideration. Thus, the example of last-chance tourism shows the influence of media coverage of climate change on the consumption system of the glacier.

In all the cases studied, the scarcity dynamic of the glacier as a resource leads the production system to substitute one object for another. At the same time, this scarcity dynamic leads the consumption system to assign a new value to the glacier as an object in relation to climate change. However, the substitution of the object by the production system poses a threat to the sustainability of glacier tourism if it involves more expensive infrastructure and luxury positioning, which could lead to this type of tourism being reserved for the elite. The analysis of the consumption and production system dynamics could be improved by better understanding the territorial context on which the site depends. This could be useful to understand the dynamics of the actors and their impact on the strategies they develop.

Management Implications

In a more general perspective, the results show that climate change and its physical impacts on mountain areas affect tourism through changes in how the object can be appropriated by stakeholders and transformed into an economic resource. The example of glacier tourism shows that a regional economic approach to the resource could be an efficient solution for designing pro-poor tourism in other areas. The results presented here will be further developed, in particular by systematically presenting the changes in production and consumption systems since the introduction of tourism on the Alpine glaciers from a geohistorical perspective. This future work will provide a better understanding of the systemic issues and consequences of the economic transformations of tourism linked to climate change. This better understanding will also allow better adaptation strategies to be developed in order to respond in a sustainable manner.

References

Apollo, M., & Wengel, Y. (2021). *Mountaineering Tourism: A Critical Perspective.* Routledge. https://doi.org/10.4324/9781003095323.

Bourdeau, P. (2018). L'après-tourisme revisité. *Via. Tourism Review*, 13. https://doi.org/10.4000/viatourism.1936.

Bussard, J., & Reynard, E. (2022). Geotourism and Interpretation of Geomorphology in Mountain Tourism. *Journal of Alpine Research*, 110–1. https://doi.org/10.4000/rga.10298.

Colletis-Wahl, K., Corpataux, J., Crevoisier, O., Pecqueur, B., Peyrache-Gadeau, V., & Kebir, L. (2008). The Territorial Economy: A General Approach in Order to Understand and Deal with Globalisation. In M. Aranguren Querejeta, C. Iturrioz Landart, & J. Wilson, *Networks, Governance and Economic Development* (21-29). Edward Elgar Publishing. https://doi.org/10.4337/9781848442689.00011.

De Gregori, T. R. (1987). Resources Are Not; They Become: An Institutional Theory. *Journal of Economic Issues*, 21(3), 1241–1263. https://doi.org/10.1080/00213624.1987.11504702.

Duvillard, P.-A., Ravanel, L., & Deline, P. (2015). Risk assessment of infrastructure destabilisation due to global warming in the high French Alps. *Journal of Alpine Research*, 103–2. https://doi.org/10.4000/rga.2896.

Groulx, M., Lemieux, C. J., Lewis, J. L., & Brown, S. (2016). Understanding consumer behaviour and adaptation planning responses to climate-driven environmental change in Canada's parks and protected areas: A climate futurescapes approach. *Journal of Environmental Planning and Management*, 60(6), 1016–1035. https://doi.org/10.1080/09640568.2016.1192024.

Gumuchian, H., & Pecqueur, B. (Eds.). (2007). *La ressource territoriale.* Economica: Anthropos.

IPCC. (2022). *Climate Change 2022: Impacts, Adaptation, and Vulnerability. Contribution of Working Group II to the Sixth Assessment Report of the Intergovernmental Panel on Climate Change* ([H.-O. Pörtner, D.C. Roberts, M. Tignor, E.S. Poloczanska, K. Mintenbeck, A. Alegría, M. Craig, S. Langsdorf, S. Löschke, V. Möller, A. Okem, B. Rama (eds.)]). Cambridge University Press. https://www.ipcc.ch/report/ar6/wg2/downloads/report/IPCC_AR6_WGII_SummaryForPolicymakers.pdf.

Jeannerat, H. (2021). Des dynamiques territoriales d'innovation aux dynamiques territoriales de valuation. *Revue d'Économie Régionale & Urbaine*, 1, 29–50. https://doi.org/10.3917/reru.211.0029.

Joutard, P. (1986). *L'invention du Mont Blanc*. Gallimard/Julliard.

Kebir, L. (2006). Ressource et développement régional, quels enjeux? *Revue d'Économie Régionale & Urbaine*,5, 701–723. https://doi.org/10.3917/reru.065.0701.

Kebir, L. (2016). Analyser les ressources et leurs dynamiques: Pour une approche institutionnelle et territoriale prenant en compte les relations producteurs-consommateurs. In B. P. Éric Glon (Ed.), *Proximités et ressources territoriales – Au cœur des territoires créatifs?* (161–172). Presses universitaires de Rennes. https://hal.archives-ouvertes.fr/hal-01456522.

Kebir, L., & Crevoisier, O. (2007). Resources development and actors coordination: What role for innovative milieus? *International Journal of Entrepreneurship and Innovation Management*, 7(2–5),204–222. https://doi.org/10.1504/IJEIM.2007.012882.

Lemelin, H., Dawson, J., Stewart, E. J., Maher, P., & Lueck, M. (2010). Last-chance tourism: The boom, doom, and gloom of visiting vanishing destinations. *Current Issues in Tourism*, 13(5), 477–493. https://doi.org/10.1080/13683500903406367.

Lemieux, C. J., Groulx, M., Halpenny, E., Stager, H., Dawson, J., Stewart, E. J., & Hvenegaard, G. T. (2018). "The End of the Ice Age?": Disappearing World Heritage and the Climate Change Communication Imperative. *Environmental Communication*, 12(5), 653–671. https://doi.org/10.1080/17524032.2017.1400454.

Mayer, M., & Abegg, B. (2022). Development of summer skiing days in Austrian glacier ski areas in the first two decades of the twenty-first century. *International Journal of Biometeorology*. https://doi.org/10.1007/s00484-022-02371-6.

Moreau, M. (2010). Visual perception of changes in a high mountain landscape: The case of the retreat of the Évettes Glacier (Haute-Maurienne, northern French Alps). *Géomorphologie: Relief, Processus, Environnement*, 16(2), 165–174. https://doi.org/10.4000/geomorphologie.7901.

Mourey, J., Ravanel, L., Lambiel, C., Strecker, J., & Piccardi, M. (2019). Access routes to high mountain huts facing climate-induced environmental changes and adaptive strategies in the Western Alps since the 1990s. *Norsk Geografisk Tidsskrift – Norwegian Journal of Geography*, 73(4), 215–228. https://doi.org/10.1080/00291951.2019.1689163.

Nesur, K., Salim, E., Girault, C., & Ravanel, L. (2022). Sharing Scientific Knowledge on Glaciers to the General Public: The Role of Glacier Interpretation Centres in Mountain Tourism Diversification Strategies. *Journal of Alpine Research*, 110–1. https://doi.org/10.4000/rga.10143.

Newsome, D., & Dowling, R. K. (2010). Setting an agenda for geotourism. In D. Newsome & R. K. Dowling (Eds.), *Geotourism: The tourism of geology and landscape.* Goodfellow Publishers Limited. https://researchrepository.murdoch.edu.au/id/eprint/2601/.

Patton, M. Q. (2002). *Qualitative research and evaluation methods.* Sage.

Purdie, H. (2013). Glacier Retreat and Tourism: Insights from New Zealand. *Mountain Research and Development*, 33(4), 463–472. https://doi.org/10.1659/MRD-JOURNAL-D-12-00073.1.

Roe, G. H., Baker, M. B., & Herla, F. (2017). Centennial glacier retreat as categorical evidence of regional climate change. *Nature Geoscience*, 10, 95–99 . https://doi.org/10.1038/ngeo2863.

Salim, E. (2023). Glacier tourism without ice: Envisioning future adaptations in a melting world. *Frontiers in Human Dynamics*, 5, 1137551.https://doi.org/10.3389/fhumd.2023.1137551.

Salim, E., Gauchon, C., & Ravanel, L. (2021). Seeing the Ice. An Overview of Alpine Glacier Tourism Sites, Between Post- and Hyper-Modernity. *Journal of Alpine Research*, 109–4. https://doi.org/10.4000/rga.8383.

Salim, E., Mabboux, L., Ravanel, L., Deline, P., & Gauchon, C. (2021). A history of tourism at the Mer de Glace: Adaptations of glacier tourism to glacier fluctuations since 1741. *Journal of Mountain Science*, 18(8), 1977–1994. https://doi.org/10.1007/s11629-021-6723-5.

Salim, E., Mayer, M., Sacher, P., & Ravanel, L. (2023). Visitors' motivations to engage in glacier tourism in the European Alps: Comparison of six sites in France,

Switzerland, and Austria. *Journal of Sustainable Tourism*, 31(6), 1373-1393. https://doi.org/10.1080/09669582.2022.2044833.

Salim, E., & Ravanel, L. (2023). Last chance to see the ice: Visitor motivation at Montenvers-Mer-de-Glace, French Alps. *Tourism Geographies*, 25(1),72-94. https://doi.org/10.1080/14616688.2020.1833971.

Salim, E., Ravanel, L., Bourdeau, P., & Deline, P. (2021). Glacier tourism and climate change: Effects, adaptations, and perspectives in the Alps. *Regional Environmental Change*, 21(4), 120. https://doi.org/10.1007/s10113-021-01849-0.

Stewart, E. J., Wilson, J., Espiner, S., Purdie, H., Lemieux, C., & Dawson, J. (2016). Implications of climate change for glacier tourism. *Tourism Geographies*, 18(4), 377–398. https://doi.org/10.1080/14616688.2016.1198416.

Wang, S., He, Y., & Song, X. (2010). Impacts of climate warming on alpine glacier tourism and adaptive measures: A case study of Baishui Glacier No. 1 in Yulong Snow Mountain, Southwestern China. *Journal of Earth Science*, 21(2), 166–178. https://doi.org/10.1007/s12583-010-0015-2.

Weber, M., Groulx, M., Lemieux, C. J., Scott, D., & Dawson, J. (2019). Balancing the dual mandate of conservation and visitor use at a Canadian world heritage site in an era of rapid climate change. *Journal of Sustainable Tourism*, 27(9), 1318–1337. https://doi.org/10.1080/09669582.2019.1620754.

Welling, J. T., Árnason, Þ., & Ólafsdottír, R. (2015). Glacier tourism: A scoping review. *Tourism Geographies*, 17(5), 635–662. https://doi.org/10.1080/14616688.2015.1084529.

Welling, J. T., Árnason, Þ., & Ólafsdóttir, R. (2020). Implications of Climate Change on Nature-Based Tourism Demand: A Segmentation Analysis of Glacier Site Visitors in Southeast Iceland. *Sustainability*, 12(13), 5338. https://doi.org/10.3390/su12135338.

Yuan, L., Lu, A., Ning, B., & He, Y. (2006). Impacts of Yulong Mountain glacier on tourism in Lijiang. *Journal of Mountain Science*, 3(1), 71–80. https://doi.org/10.1007/s11629-006-0071-3.

Part IV

Concluding Remarks

16 Pro-poor Mountain Tourism-Through a Journalist's Eye

Prerna Prasad

During one of my first visits to the mountains, I asked somebody from the village, "Where can we go for trekking?" The villager asked, "What is trekking?" We explained it to him and he replied, we do it every day, all the time. "Trekking for us is not a sport but a part of life." That changed my perspective and opened my subtle consciousness to work for sustainable tourism.

There are three ways of looking at mountain tourism. One, in places like Switzerland, where everything is developed, it is process-oriented, well-regularised by its government and operated by the private sector. The second one is at places like Shimla in India where it has unevenly grown, with a lack of regularity and is exploiting the environment. And the third one is at places like Kyrgyzstan where mountain tourism is yet on the path of exploration as a source of revenue.

Here is an observation shared by a friend cum avid traveller Alka Kaushik: "While Switzerland looks like a manicured garden, where everything is regulated, process-oriented, domesticated, countries like India are raw, fresh, unregulated and nature is still in its original form at many remote places." What is better and what is not, is a topic of discussion, but something which all would agree upon, is the practice of sustainable tourism.

Community-Owned Tourism

Tirthan Valley is a destination on the river Tirthan in Himachal Pradesh, India. Raju Bharti with the support of a few local villagers filed a case in the court against the building of a hydroelectric power project on the river Tirthan. After a long-drawn battle of five years, he saved the river from construction of the dam. Tirthan is one of the few rivers in India on which a hydroelectric project has not been built. Raju Bharti is also involved in mountain tourism. He has a quaint guest house run by his family. His sons are also involved in camping and other adventure activities. Raju Bharti and his family's efforts need to be highlighted as they are the only ones who have added more trees to their compound than concrete structures.

Most of the other villagers have leased their land to outsiders which has led to the construction of big hotels and restaurants. The greed of most village

DOI: 10.4324/9781003392583-20

folks has led to the exploitation of the land, while the positivity and grit of one man has made him hold his fort. Raju Bharti's family is still known for their hospitality and are the best local guides for mountain tourism in Great Himalayan National Park, which is a UNESCO Heritage site.

Ownership of the land by locals is most important for promoting pro-poor mountain tourism. The locals must not sell or lease out their land in the greed of a meagre sum of money.

A big success story of community-owned tourism is that of the Sherpas in Nepal. Sherpas are the original inhabitants of the area around Everest base camp. When mountain tourism opened up in Nepal, especially after Sir Edmund Hillary & Sherpa Tenzing Norgay climbed Mount Everest, a lot of job opportunities were created for the Sherpa community. The Sherpas worked as porters and guides, opened guesthouses, and provided food and transportation facilities. Today, the Sherpas are the richest community in Nepal and have benefited the most economically from mountain tourism.

Mountaineering is a Mind Game

Gangotri Soneji, an avid mountain climber, successfully undertook the Trans-Himalayan expedition at the age of 63 despite having arthritis. When asked what challenges she faced as a women climber, Gangotri narrates, "Men are more strong physically while women are more strong mentally. And mountain climbing is more of a mind game." So, the mental strength of women outweighs the physical strength of men.

Mountains are Magic

Josiah Waithaka, a mountain guide at Mount Kilimanjaro and Mount Kenya, shares an interesting fact. He says that, by climbing the mountain, he can experience winters even during the summer season. Many of you might be thinking how! Mount Kilimanjaro is the world's largest free-standing mountain. When you start climbing, it is warm and as you move up you experience the glaciers and the temperature goes down.

It is interesting to know that Mount Kenya in Africa passes through the equator, and still has around 2 hours of glacier hike. Isn't it magical? Every mountain has a vibe and it is much more than an adventure sport.

Mountains are Sacred

Most people recognise mountains with adventure sports or mountaineering expeditions. Mountains are not just for climbing, but they are "sacred", especially for the locals. In the words of Serku Sherpa, who has authored a book on "The Sherpas and Their Original Identity", Mountains are the protector for the Sherpa community. He says that Sherpas are mostly Buddhists and, as per their belief, humans are guests on planet Earth. He says, "If we do not respect planet Earth, nature will take action against us."

Serku Sherpa shares an interesting practice of his community. He says, "In Nepal, mantras (sacred words) are written on stone, so nobody touches that stone for generations as it is considered divine. Scientifically, these stones protect the land from soil erosion." So, every spiritual practice has a science behind it. Hence, mountains are not mere stones but have a strong spiritual value that must be respected and preserved.

In the words of Gangotri, an avid climber, "mountain climbing is about connecting with the inner self. Mountain climbing is not for fun. Today, Everest base camp has TV, snooker, and lot of other luxury facilities." She further adds, "If you want to party and watch TV, go to Singapore, Dubai, Los Angeles, Mumbai. Don't go to the mountains to party. Enjoy nature and leave it as it is for the future generations!"

Where to Learn From?

Uzbek tourism expert Michael Shamshidov, who is working tirelessly on promoting Central Asian countries on the global tourism map, shared an interesting anecdote. Michael mentioned that he pushes his government to send its officials to learn from India or other Asian countries rather than Germany or European nations. He adds that the people of his country have a mindset similar to that of Asian countries. Asian nations like India have been dealing with similar problems and have coped with them in recent years – in contrast to a country like Germany which has passed through those challenges 100 years back.

What is Poverty, Really?

From my experience of travelling the length and breadth of India, I realise that the people in the villages are living a more luxurious life, as the real luxury lies in access to fresh air, access to pure water and access to food that is chemical-free.

The New Luxury

People residing in mountainous areas have access to fresh mountain air, pure water from springs and chemical-free food grown in their own backyards. The developed countries need to recognise this new luxury and encourage it.

Many people from metro cities are gradually shifting back to villages as they have recognised this new luxury. As a part of responsible tourism, it is also our responsibility to promote reverse migration. In a recent example, one of our eco-hotels based in Jageshwar in Uttarakhand (Himalayan belt), motivated almost 7 members of a family to migrate back to their village by helping them start tourism activities in the mountainous areas.

Vocal for Local

When we speak about sustainability, we refer to people, planet and profit. The profit here refers to the revenue generated by the local community. In developing or underdeveloped countries, the local community consists predominantly of economically weaker people. So, any tourism activity that is devoid of benefitting the poor is by nature non-sustainable.

To benefit the poor who live in the mountain areas, here are some key measures to be practised by travellers:

- When booking your accommodation, look for homestays or community-owned accommodations.
- While having your meals, go for local cuisines prepared by the local chefs. Pizza and pasta can be consumed when you are holidaying in Italy.
- While going sightseeing, always go on an itinerary prepared by local tour operators as they have the best knowledge of their area.
- Mountain areas offer a lot of adventure activities like skiing, mountain biking, trekking, camping, nature walks, birding, and many more. While choosing your guide, prefer a local. This may require a little due diligence.

As tourists, we do a lot of research when it comes to the type of accommodation, amenities, budget and much else. If we can add a little attention to who owns the company or organisation, it would benefit the local economy in the real sense. Also, it goes without saying, that when you buy anything, be it souvenirs for your friends or shop for yourself, go for local handmade products.

Natural Buildings

The buildings that have survived since ages are all made of natural materials. This includes the Taj Mahal, Egyptian pyramids, Stonehenge or any other. Moreover, concrete is one of the largest pollutants used in construction. The major mountain destinations in India, including Shimla, Manali, Ladakh etc., have been converted into concrete jungles. So, my venture Ecoplore, a Sustainable Tourism initiative, has decided to promote only hotels that are made of natural materials like mud, bamboo, stone or any other local architecture. All these eco-hotels have a minimum of 33% tree coverage inside their compounds.

Cap on the Number of Tourists

This is something which needs to be practiced more widely. It refers to putting a cap on the number of tourists travelling to a particular tourist destination, so that the pressure on the destination can be reduced. For instance, in India, there is a cap on the number of tourists entering Rohtang Pass. Rohtang Pass connects Manali to the Himalayan glaciers. Vehicles entering this area need permission. On any day, only 1000 vehicles are permitted to enter

Rohtang Pass. The permission is provided on a first-come-first-served basis. The beautiful Rohtang Pass is one of the best-preserved areas with multiple waterfalls on the way. This is the best way to keep nature intact and to not add to its exploitation. In many countries, tourist destinations are also closed down for a certain period to give them time to heal. This is another way of balancing tourism activities along with earning revenue.

Ban Luxurious Helicopter Tours in Mountain Areas

The use of helicopters in mountain tourism has a terrible impact and can become a community killer – as Michal Apollo mentioned. The noise created by the helicopters during take-off and landing disturbs the wildlife around the area. Also, due to the air movement caused by the helicopters, the mountain glaciers are melting faster. Helicopter tourism has also taken away the jobs of many poor people living in the mountainous areas. Helicopters must be used for rescue and medical emergency, but tourism by helicopter must be stopped in the best interests of the planet and the poor.

Nature Conservation

Nature-based tourism can only be sustained through environmental conservation. Since mountain tourism is dependent on beautiful mountain views, scenic rivers, waterfalls, mountain springs, forests, and wildlife, preserving all of these is equally important. Else the destination itself may lose its significance. Mussoorie in India is a perfect example of this.

Mussoorie was a very popular mountain destination in India. At one point in time, it was flocked by tourists to see its beautiful landscape and experience the exquisite snowfall. There used to be regular snowfall in Mussoorie until 19 years back. People from far and wide travelled to see the scenic beauty of this coveted hill station, also referred to as the "Queen of the Hills". Gradually, due to unplanned and unregulated tourism, there has been a constant rise in concrete structures by cutting mountains and trees. This has led to an increase in temperature, reducing and almost ending snowfall to the disappointment of visitors. As a result, Mussoorie is no longer on the bucket list for avid travellers. Many new destinations are emerging, but if we continue to grow the greed, even those may share the fate of Mussoorie.

While developing mountain tourism, the government needs to factor in the geographical terrain of the area. For instance, four-lane roads are being built in India by cutting mountains that are prone to landslides. Developing mountain tourism at the cost of natural calamities is a mindless way to go forward.

While speaking to Gangotri Soneji who has been climbing Himalayan mountains for the last 40 years, she mentioned that 40 years back there used to be mountain springs every one hour of the climb but now they must walk for a minimum of 8 to 9 hours to find a water source. Most mountain springs have dried up leading to water scarcity.

Environmentalist Sonam Wangchuk from India's Ladakh region has been sitting on fast to raise people's awareness of climate change. Known for its snow-clad mountains, Ladakh has not seen any snow in the winter months of October 2023 to February 2024. This is alarming. Climate change is real, and its impact is highest in fragile environments, especially on the poor living in the mountain areas. As Sonam Wangchuk wisely says, people in cities have to reduce their consumption and practice a more sustainable lifestyle so that poor people in the mountains can survive and live peacefully.

Carbon Footprint by Tourism Industry and Grassroots Initiatives

The tourism industry contributes almost 10% of carbon emissions worldwide. And, as an individual or organisation, our aim should be to reduce these. I am the founder of Ecoplore, which is an aggregator of eco-friendly hotels in India. We promote only those hotels that are made of mud, wood, bamboo, stone or any other local materials. All our eco-hotels maintain a minimum of 33% green coverage inside their compounds. They also grow their own vegetables, have fruit trees, maintain rainwater harvesting and solar panels, do waste management, and follow several other sustainability measures. Ecoplore has 1250+ such eco-hotels across India. We have had the carbon emission of our eco-hotels calculated by a Europe-based agency "Know Carbon". We have thereby found out that, while a luxury hotel emits 1000 kg of carbon per room per night, our eco-hotels emit just 9 kg of carbon per room per night.

Mountain Tourism and Sustainability

For many tourists, travelling to mountains is about enjoying the fresh air, participating in adventure activities, clicking beautiful pictures and uploading them on social media – and the worst of all is excessive drinking of alcohol & playing loud music in the forests. Such activities create all kinds of pollution, air pollution by travelling in private cars, soil pollution by staying in hotels made of concrete, water pollution by throwing plastic waste (chips packets & pet bottles) in waterfalls & rivers, sound pollution by playing loud music at night. So, what should be done?

Here are some hacks that tour operators should practice while conducting mountain-based tours

- Take small groups for trekking or other activities.
- Follow the existing trail for the trek.
- Do not allow guests to play loud music at night.
- Offer healthy and home-made snacks to the guests so that they don't have to go for fast food stored in plastic packets.
- Many eco-hotels ask the guests to take back the plastic they have brought with them.

- Refillable toiletries can be used instead of single use soaps and shampoos.
- Please offer water in glass bottles instead of plastic.
- Food must be served in steel or ceramic utensils.
- No use of disposables (single use plastic).

Along with these, there are certain measures that hotel owners should take to make their business more sustainable. They should avoid new construction and instead renovate a heritage buildings to turn it into an eco-hotel. If a new building must be built, it should be with natural materials such as mud, wood, bamboo or stone. Concrete buildings should be strictly avoided. The hotels must deploy measures like rainwater harvesting, solar panels, grey-water treatment, waste segregation, composting, etc. Growing vegetables and having fruit trees is encouraged. My venture "Ecoplore" aggregates all such eco-hotels that give a unique experience to their guests and at the same time add to the conservation of the planet.

It is not just the hotels but the travellers' role that becomes equally important in contributing to sustainable tourism. Here is what we as travellers can do to benefit pro-poor mountain tourism:

- Carry your own water bottles that can be refilled. Do not buy or ask for plastic bottles from the hotels.
- Carry your cloth or jute bags for shopping.
- Buy local handmade products as souvenirs.
- Choose public transport like trains and busses instead of private cars. Avoid taking airplanes as much as possible. Flights have the maximum carbon emission.
- Carry your own cutlery instead of using disposables during your travel.
- Avoid buying chips packets and cold drinks. Go for healthy snacks like fruits and homemade snacks.
- While booking the hotel, choose an eco-friendly one constructed from natural building materials like mud, wood, bamboo, stone or any other local architecture.
- Carry your own towels and toiletries.
- Eat local cuisines and seasonal stuff.
- Choose local guides for activities like trekking, nature walking, birding and others.
- Respect local culture and traditions. For example, when visiting a Gurudwara (Sikh temple) you are expected to cover your head.
- Avoid littering and throwing garbage on the mountains. Littering is a major issue in the mountains. Even the Everest base camp has been dumped with plastic. If you are carrying plastic, please bring it back with you and dispose it in the municipal corporation bins in the cities.
- Avoid playing loud music in the forests and do not disturb the wildlife.

Slow Travel/Tourism Concept and the Final Remark

Slow tourism is an alternative tourism choice in contrast to mass tourism. It refers to going on a longer vacation rather than for just one or two nights, which is the key to traveling better, smarter, and more passionately. One must go for long travel instead of weekend getaways. The carbon emission due to frequent travel in weekend getaways is very high. During slow travel, emissions reduce as we stay at one place for a longer time and the transportation is less. Slow travel also leads to an understanding of culture, lifestyle, food, and many other things at the destination we are visiting. The concept of slow travel along with sustainability is the need of the hour to advance pro-poor mountain tourism.

I spent my childhood in a remote tribal area of Chattisgarh, Central India, and my heart belongs to the tribal people. I would like to end my chapter with this word of wisdom I strongly believe in: the indigenous tribes all across the world are the best conservationists. They take all their decisions considering their impact on future generations. By contrast, we, who call ourselves “modern and civilised”, mostly decide by considering our present and especially our own comfort and convenience. Due to climate change, I have personally decided not to give birth to a human, as their future is in jeopardy. I feel pity looking at the kids, especially those born during COVID-19 times, who barely have access to clean air, water and food. The future age of technology, devoid of nature is dangerous. The clock is ticking, and we have it in our power to adopt a sustainable lifestyle and preserve our planet in the interests of all!

17 The Future Outlook on Poverty Reduction Strategies with the Use of Mountain tourism

Yana Wengel, Michal Apollo and Thomas Pogge

Why This Book Was Written

This book offers examples of pro-poor tourism in mountain areas, which has emerged as a multifaceted approach to promoting sustainable development while addressing the needs and aspirations of marginalised communities living in these picturesque landscapes. Many mountain communities face challenges related to poverty and economic marginalisation. Limited access to markets, arable land, and employment opportunities can contribute to lower household incomes and higher poverty rates compared to lowland areas. Poverty in mountain regions can be exacerbated by factors such as geographical isolation, rugged terrain, and dependency on weather-dependent livelihoods. Due to geographical isolation, inadequate infrastructure, and limited government investment, mountain communities find it hard to access essential services such as healthcare, education, clean water, and sanitation. Providing services to remote and dispersed populations in mountainous areas is more costly and logistically challenging. Tourism can improve economic conditions in such locations by generating revenue, creating job opportunities, and improving infrastructure.

In recent decades, nature-based tourism, adventure tourism, and mountain tourism have experienced a surge in popularity, especially following the COVID-19 pandemic. Nature is renowned for offering both physical and mental health benefits, thereby contributing significantly to the overall well-being of visitors. Mountains have increasingly become popular tourist destinations for people seeking adventure, tranquillity, and cultural immersion in mountain landscapes characterised by their unique ecosystems and cultural diversity. Tourism in mountain areas can play a significant role in the socio-economic development of local communities, providing opportunities for income generation, employment, cultural exchange, and infrastructure development. However, the benefits of tourism in mountain regions have not always been equitably distributed among the locals, often intensifying socio-economic disparities and environmental degradation.

DOI: 10.4324/9781003392583-21

To Manage, You Need a Plan: All Stakeholders are on Board!

Robust, thoughtful, but also socially fair change cannot be achieved without the full participation of all stakeholders in the planning of change and its implementation. In our recent book, *Mountaineering Tourism*, we published a model that, based on other work, described in simple terms a scheme for all stakeholders to work together to achieve sustainable development. Fig. 17.1 shows this model after minor modifications. It can serve as a basic model for bringing together all stakeholders and jointly developing guidelines for devising measures that comply with the pro-poor principle.

The management of mountain areas should be under the constant control of representatives of each group concerned. Managers should take action in the order and with the concept outlined below, as these steps will only be effective if coordinated. A coordinated approach is essential to ensure that strategies align with the overarching goal of sustainable development. This involves a multifaceted approach as outlined below.

Firstly, detailed guidelines must be developed, including comprehensive protection plans and regulations. However, introducing such measures should be preceded by thorough consultations with all relevant stakeholders. The collaborative process helps ensure that changes do not treat unfairly any

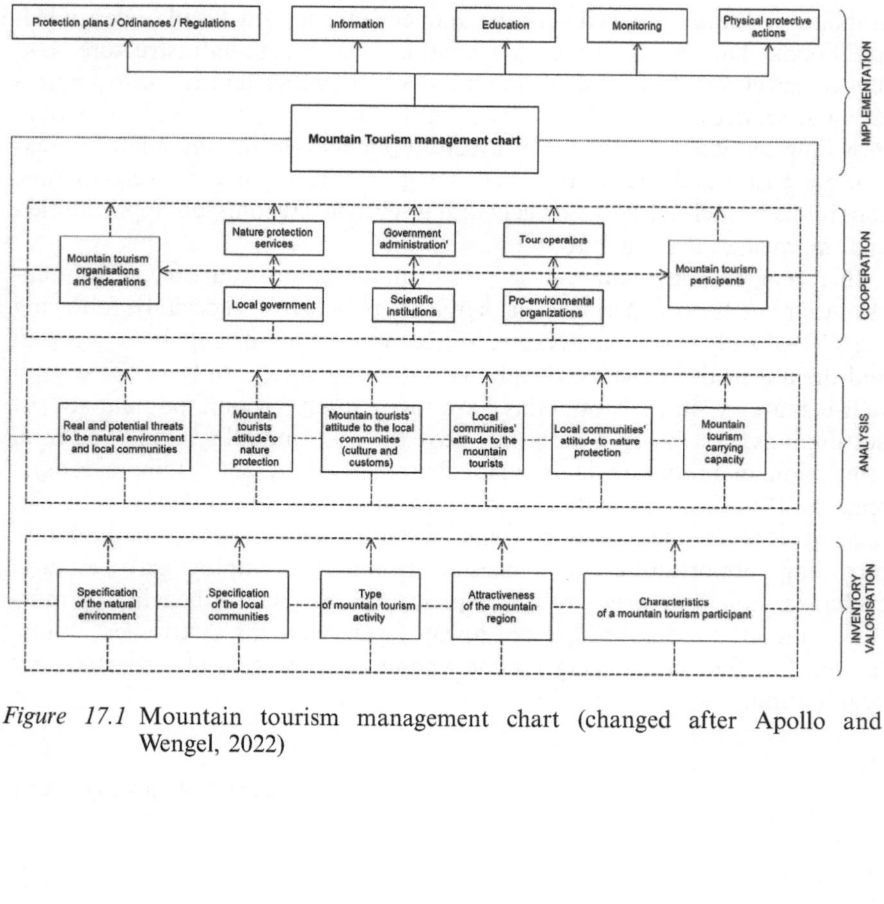

Figure 17.1 Mountain tourism management chart (changed after Apollo and Wengel, 2022)

single party involved. Secondly, clear and coherent communication of these guidelines is paramount. Information regarding conduct standards and consequences for non-compliance should be widely disseminated through various channels, such as websites, signage, and guides. Proper communication strategy ensures that all individuals engaging in mountain tourism are well-informed and understand their responsibilities. Moreover, ongoing education initiatives are crucial for fostering understanding among all parties involved. Just providing information may not suffice; educational efforts and capacity development are needed to convey the rationale behind the pursued policies with their potential benefits and drawbacks. Monitoring mountain areas is also essential to detect and address deviations from the established plan. Rapid improvements in one management aspect may inadvertently lead to negative effects elsewhere. Therefore, timely intervention is necessary to maintain balance and harmony among the ecosystem, community, and other stakeholders. Finally, physical protective measures may need to be implemented as a last resort. These could include trail improvements and measures to regulate tourist numbers in light of the area's environmental and social carrying capacity (see e.g.: Apollo 2021; Apollo & Andreychouk, 2022).

Recognising the views and position of each stakeholder ensures that the use of tourism to fight inequality and poverty is carried out voluntarily, as the interests of each party are taken into account. It is essential to recognise that management in pro-poor mountain tourism is not solely focused on restricting or halting development. Instead, it aims to facilitate rational progress that benefits all parties involved while preserving the mountain environments' unique natural and cultural heritage. Ultimately, the aim of pro-poor mountain tourism management is to create opportunities for economic growth and poverty alleviation among local populations. Through collaborative efforts and a focus on equitable development, mountain tourism can become a powerful tool for positive change in the lives of those living in mountain regions.

Implications for Managing Mountain Tourism for Poverty Alleviation

Based on the management implications recommended by the authors of the various chapters included in this book, we have identified the main training grounds for action. Interestingly, although the mountain areas described are located in different parts of the world and are inhabited by different populations (race, faith, culture), it was not difficult to combine these narratives. Such harmonisation underscores the universal nature of the challenges and of the proposed solutions across different mountainous regions. The commonality stems from the distinctive attributes of mountain environments, which mould their inhabitants and consequently give rise to similar issues.

While pro-poor tourism is generally seen as a positive approach to tourism development, critiques have emerged regarding its implementation and effectiveness. Opponents argue that initiatives often involve superficial community participation, risking tokenism and marginalisation. To combat this, genuine engagement with local communities in planning and benefit-sharing is crucial. Concerns also arise about dependency on tourism, prompting the need for diversified livelihood options beyond tourism. Additionally, there is a risk of cultural commodification and inequitable benefit distribution within a community, demanding authentic cultural interactions and transparent benefit-sharing mechanisms. Environmental impacts pose another challenge, urging sustainable tourism practices to minimise negative effects. Managing these challenges requires a collaborative approach that prioritises community involvement, community well-being, sustainable development, and genuine stakeholder partnerships that foster positive tourism.

Community involvement and ownership are critical principles of pro-poor tourism. By actively engaging local residents in planning and decision-making, tourism initiatives can be tailored to meet the needs and aspirations of the communities they aim to benefit. This participatory approach ensures that tourism development aligns with local values and priorities and empowers communities to become active stakeholders in their development. Moreover, pro-poor tourism emphasises the creation of economic opportunities for marginalised groups, including women, indigenous peoples, and rural populations. Tourism can generate income and employment opportunities contributing to poverty reduction and social inclusion by promoting community-based enterprises such as homestays, handicrafts, and guided tours. In addition to socio-economic benefits, pro-poor tourism in mountain areas strongly emphasises environmental sustainability. In this book, editors and authors advocate for sustainable tourism practices that are designed to minimise negative impacts on natural resources and to promote conservation efforts, recognizing that mountain ecosystems are vulnerable to environmental degradation and climate change.

This book frames pro-poor tourism in mountain areas as a holistic approach that seeks to harness the potential of tourism as a vehicle for positive change, ensuring that the benefits are shared equitably among all members of society while safeguarding the cultural and ecological integrity of these pristine landscapes. Part one of this book presents various strategies for activating pro-poor mountain tourism; part two highlights the impacts and challenges associated with developing pro-poor tourism in mountain areas; part three provides some lessons relevant to pro-poor tourism in mountain areas from advanced economies. This book offers a critical perspective on mountain tourism and its potential for sustainable development and for addressing poverty and inequality in particular. In this concluding chapter, we also wish to highlight that pro-poor mountain tourism is a multifaceted phenomenon; hence, tourism planning in mountain areas must encompass socio-cultural, political, economic, and environmental dimensions. Based on the contributions of all authors, the management implications for pro-poor mountain tourism merged into four key themes, each emphasising crucial aspects of sustainable and equitable development in mountain regions.

One of the central themes that runs like a common thread throughout the book is **community empowerment and participation**. Vital to the success of pro-poor mountain tourism development is the active involvement of local communities. Integrating local natural and cultural assets into tourism planning fosters a sense of ownership and ensures that development initiatives align with community needs and aspirations. Encouraging community participation, promoting community-based tourism models, and investing in local conservation efforts are essential strategies for empowering mountain communities. Another central theme concerns **governance and stakeholder collaboration**. Collaboration among stakeholders, including government bodies, private sector stakeholders, NGOs, and community organisations, is crucial for effectively implementing pro-poor tourism initiatives. Government support through policy frameworks and infrastructure development is crucial in creating an enabling environment for sustainable tourism. Transparent governance structures facilitate inclusive decision-making processes and ensure that tourism benefits are distributed equitably among all stakeholders. Collaboration among stakeholders is critical for addressing social, environmental, and economic impacts, focusing on aligning tourism offerings with local economies and preserving cultural and environmental heritage. In many regions, local communities have an optimistic attitude toward tourism as they understand the positive economic impacts it can have. But remote communities often lack the skills required to implement ideas as a business. **Capacity building** initiatives can provide local residents with the necessary tools. It is crucial to educate communities on starting up businesses and promoting their destinations while emphasising the uniqueness of their culture and environment. Developing the capacity of individuals is essential for maximising the potential of mountain tourism while ensuring its long-term sustainability. A fourth important theme in this book is **climate change adaptation and resilience**. Climate change poses significant challenges to mountain tourism, altering natural environments and disrupting traditional livelihoods. Adapting to these changes requires innovative strategies that promote resilience and sustainability. Emphasising regional economic approaches and conducting research to understand the systemic impacts of climate change on tourism are essential steps toward developing effective adaptation strategies. Engaging local communities in climate-sensitive tourism activities linked to the local economy enhances resilience and fosters poverty eradication.

The future of mountain tourism is likely to be shaped by a combination of factors, including evolving travel trends, technological advancements, and efforts to mitigate climate change. Sustainable and responsible tourism practices are expected to become increasingly important. Tourism products focusing on conserving fragile mountain ecosystems, preserving local cultures and traditions, and benefiting local communities should become mainstream and not represent an exception. Additionally, there may be a shift towards more immersive and experiential tourism offerings, catering to travellers seeking authentic and meaningful experiences in natural settings. However, challenges such as over-tourism, environmental degradation, and socio-economic disparities must be addressed through collaborative efforts among stakeholders, including governments, local communities, tourism operators, and conservation organisations. At the same time, tourists themselves will have to adopt a responsible attitude towards host communities and the destinations. Responsible tourists show respect towards customs and traditions, support local businesses, and aim to minimise their environmental impact. Such tourists contribute positively to the community and educate themselves and others about social, environmental, and economic issues at the host destination to ensure that tourism benefits host communities and destinations while minimising negative impacts.

Ultimately, the future of mountain tourism will depend on striking the right balance between the economic benefits of tourism and the need to protect and sustainably manage mountain environments for future generations. The future prospects for poverty reduction strategies using mountain tourism are promising, with opportunities to leverage the economic benefits of tourism to uplift marginalised communities in mountainous regions. However, success will depend on implementing sustainable tourism practices, fostering genuine community involvement, and addressing challenges such as environmental conservation, cultural preservation, and equitable benefit distribution. Collaborative efforts among stakeholders, including governments, local communities, and tourism operators, will be essential for maximising the positive impacts of mountain tourism on poverty reduction while minimising potential drawbacks.

Fair Mountain Tourism: Future Action Required

As mountain tourism continues to expand, there is a growing recognition of the need for sustainable and equitable practices within this industry. The present volume illustrates that pro-poor mountain tourism emerges as a response, aiming to benefit local communities economically while preserving mountain regions' natural and cultural integrity. Achieving fairness in mountain tourism brings in additional factors, including climate change impacts, gender dynamics, and profit distribution.

Climate change poses significant challenges to mountain ecosystems, affecting biodiversity, water resources, and local livelihoods. Mountain areas are characterised by rich biological and cultural diversity and offer essential resources like water, food, and energy to communities in their region and beyond. Mountains provide freshwater for half the global population and hold immense biodiversity and cultural significance. However, mountain regions face serious challenges due to climate change, including rising temperatures, melting glaciers, and more frequent natural disasters. Mountain communities, while resilient, are vulnerable due to their fragile environments and socio-economic circumstances. They typically lack the infrastructure to adapt to worsening conditions that can threaten livelihoods and lead to conflicts and displacement. Climate impacts in the mountains spread beyond their borders, affecting lowlands and regions worldwide. As temperatures rise, hazards like floods and wildfires intensify, causing damage to infrastructure and lives both locally and downstream.

Changing seasonal patterns also impact mountain tourism, posing risks to winter recreation and mountaineering activities.

Despite such challenges, evidence points to opportunities for effective responses to climate risks in the mountains – though adaptation solutions depend on socio-economic and climatic factors, and stakeholders prioritise different dimensions of risks (López et al., 2017; McDowell et al., 2020). A growing body of research indicates that women's close connection to natural resources (water, agriculture, forest resources etc.), stemming from their reliance on them, enhances the success of their conservation efforts, which is evidenced in many locations worldwide (Aijazi & Basu, 2021).

All this suggests that a participatory approach to problem-solving, engagement of all members of local communities, and cooperation among stakeholders can help identify climate stressors and facilitate better adaptation to climate change. In this context, fostering collaborative initiatives among stakeholders, such as governments, local communities, and tourism operators, is pivotal since networking and participatory problem-solving are likely to improve the efficiency of suggested measures. Furthermore, pro-poor mountain tourism strategies must integrate climate resilience measures to mitigate these impacts and ensure the long-term sustainability of tourism-based economies.

Gender dynamics in mountainous regions have garnered increasing attention (e.g. chapter seven; Apollo et al., 2023; Hall et al., 2023). In the past, women in mountainous regions often lacked access to education, typically adhering to traditional roles within their communities. Moser (1993) emphasises women's multifaceted contributions and highlights that the roles in daily life include reproductive work and childcare, productive work (both on- and off-farm), and community management and politics. In their daily lives, women are more likely than men to engage in agricultural activities and often remain in rural areas while men migrate to urban centres.

Women play a crucial role in environmental stewardship within mountain ecosystems, contributing significantly to resource management, biodiversity conservation, and food and water security. Due to their close relationship with mountain environments, women are especially attuned to environmental changes and degradation. For example, in the communities involved in tourism, women tend to have a better understanding of the complex relationship between their homestay and the environment and its resources; they know how safeguarding mountain ecosystems is crucial for the success of their enterprises. Such awareness increases women's involvement in conservation initiatives (Aijazi & Basu, 2021). This evolving dynamic emphasises viewing mountain women as essential for sustainable mountain development, particularly in addressing climate change-related challenges (Rudaz & Debarbieux, 2012; Nellemann et al., 2011).

Furthermore, in a significant shift, girls now have education opportunities and are encouraged to pursue broader roles. This shift has transformed women's physical, material, and social environment in mountain regions. While agriculture and related activities have historically defined the lives of mountain women, the changing circumstances have led to more complex and diverse realities. For example, in Nepal, initiatives such as promoting women-owned endeavours, such as guest houses or tourist guide services, empower women economically, address their resource constraints, energise them, and enhance their participation in community life. Entrepreneurship is a potential tool for socio-economic empowerment, especially for women. But women in mountainous areas also encounter difficulties, such as restricted access to services and communication networks. Prioritising education and financial literacy are essential to overcoming these challenges. Establishing entrepreneurial networks among mountain communities can create opportunities to sustain women's enterprises. Empowerment and balance in gender dynamics are crucial in pro-poor mountain tourism, with women often playing central roles in community-based tourism initiatives. Recognizing and facilitating women's contributions can enhance the effectiveness and inclusivity of pro-poor tourism efforts in mountain regions.

Historically, many mountain communities have seen profits from resource extraction (such as timber, minerals, and hydropower), but these have not been returned to them, leading to economic disparities (Perlik, 2015). In 1995, the UN Commission on Sustainable Development acknowledged the importance of reevaluating resource flows and the need for equitable distribution of benefits to local communities in mountain areas. Unless mountain communities are empowered with secure ownership and usage rights, access to information, participation in decisions, and improved economic standards, they risk exploiting short-term opportunities and depleting crucial resources (Byers, 1995). Without adequate empowerment and control, mountain communities may be unable to prevent overexploitation by outside parties, highlighting the need for their inclusion in decision-making processes and resource management (Apollo & Wengel, 2022). Despite potential adverse impacts, tourism diversifies mountain economies and serves as a vital source of income and economic prospects for mountain communities (Romeo et al., 2021). Profit distribution is key to ensuring that the benefits of pro-poor mountain tourism reach local communities equitably. To ensure fair profit distribution, especially from tourism, mountain communities must engage all stakeholders and implement transparent management and communication policies. Fair distribution mechanisms, such as community-based enterprises and revenue-sharing agreements, are essential for fostering economic development and reducing poverty in mountain areas.

Future research in pro-poor mountain tourism should explore innovative strategies for enhancing community participation and ownership, assess the socio-economic impacts of tourism on local populations, investigate alternative tourism models that prioritize sustainability and equity, and develop robust monitoring and evaluation frameworks for measuring the effectiveness of pro-poor tourism initiatives over time.

References

Aijazi, O. and Basu, S. (eds.) (2021). *Critical approaches to gender in mountain ecosystems.* IUCN.

Apollo, M. (2021). *Environmental impacts of mountaineering: A conceptual framework*. Springer. https://doi.org/10.1007/978-3-030-72667-6.

Apollo, M., Mostowska, J., Legut, A., Maciuk, K., & Timothy, D. J. (2023). Gender differences in competitive adventure sports tourism. *Journal of Outdoor Recreation and Tourism*, 42, 100604. https://doi.org/10.1016/j.jort.2022.100604.

Apollo, M., & Andreychouk, V. (2022). *Mountaineering Adventure Tourism and Local Communities: Social, Environmental and Economics Interactions*. Edward Elgar Publishing. https://doi.org/10.4337/9781802209389.

Apollo, M., & Wengel, Y. (2022). *Mountaineering tourism: A critical perspective*. Routledge. https://doi.org/10.4324/9781003095323.

Byers, E. 1995. *Mountain Agenda: environmentally sustainable and equitable development opportunities*. Background paper in support of the global NGO Consultation on the Mountain Agenda. Lima, Peru.

Hall, J., Boocock, E., & Avner, Z. (Eds.). (2023). *Gender, Politics and Change in Mountaineering: Moving Mountains*. Springer. https://doi.org/10.1007/978-3-031-29945-2.

Lopez, S., Jung, J.-K., & López Sandoval, M. (2017). A hybrid-epistemological approach to climate change research: Linking scientific and smallholder knowledge systems in the Ecuadorian Andes. *Anthropocene*, 17, 30-45. https://doi.org/10.1016/j.ancene.2017.01.001.

McDowell, G., Harris, L., Koppes, M., Price, M. F., Chan, K. M. A., & Lama, D. G. (2020). From needs to actions: prospects for planned adaptations in high mountain communities. *Climatic Change*, 163(2), 953–972. https://doi.org/10.1007/s10584-020-02920-1.

Moser, C.O.N. (1993). *Gender Planning and Development: Theory*, Practice and Training. London and New York: Routledge.

Nellemann, C., Verma, R. and Hislop, L. (eds.) (2011). *Women at the Frontline of Climate Change: Gender Risks and Hopes. A Rapid Response Assessment*. Birkeland Trykkeri AS.

Perlik, M. (2015). Mountains as Global Suppliers: New Forms of Disparities Between Mountain Areas and Metropolitan Hubs. *Journal of Alpine Research*, 103(3). https://doi.org/10.4000/rga.3142.

Romeo R., Russo, L., Parisi F., Notarianni M., Manuelli S. and Carvao S., UNWTO. (2021). Mountain tourism – Towards a more sustainable path. FAO. https://doi.org/10.4060/cb7884en.

Rudaz, G. and Debarbieux, B. (2012). Mountain Women: Silent Contributors to the Global Agenda for Sustainable Mountain Development. *Gender, Place and Culture* 19(5), 615–634. https://doi.org/10.1080/0966369x.2011.625084.

Index

Note: *Italic* page numbers refer to *figures*; **Bold** page numbers refer to **tables**